T0250447

Medical Device and Equipment Design

Usability Engineering and Ergonomics

Medical Device and Equipment Design

Usability Engineering and Ergonomics

Michael E. Wiklund

Taylor & Francis
Taylor & Francis Group

Boca Raton   London   New York   Singapore

A CRC title, part of the Taylor & Francis imprint, a member of the
Taylor & Francis Group, the academic division of T&F Informa plc.

Published in 1995 by
CRC Press
Taylor & Francis Group
6000 Broken Sound Parkway NW, Suite 300
Boca Raton, FL 33487-2742

No claim to original U.S. Government works
Printed in the United States of America on acid-free paper
10 9 8 7 6 5 4 3 2 1

International Standard Book Number-10: 0-935184-69-4 (Hardcover)
International Standard Book Number-13: 978-0-935184-69-3 (Hardcover)

Library of Congress Cataloging-in-Publication Data

Catalog record is available from the Library of Congress

Contents

Section 4 Designing the User Interface 113

Preface

Does the world need a book that focuses on making medical devices easier to use by improving their design? Obviously, I think the answer is "yes." Consider the fact that many medical devices directly or indirectly serve a lifesaving, life-preserving, or life-enhancing purpose. User-interface design deficiencies and outright flaws can interfere with these purposes. Sometimes, the result simply may be reduced medical worker productivity or frustration. At other times, however, the result may be serious user errors—mistakes or mental slips—that can lead to patient injury or even death. In short, usability concerns need to be raised to the same level as traditional technological, manufacturing, and economic concerns.

Many good examples of design can be found, and most manufacturers care deeply about users' needs. At the same time, however, many companies need to convert their concern for users into action by taking advantage of state-of-the-art user-interface design methods. They need to invest in a more systematic approach to user-interface development that takes full advantage of the growing base of knowledge about human beings and human performance. I hope this book will serve as both a motivator and a design resource for such companies. It also can provide nonspecialists working in small design departments sufficient orientation to the user-centered design process to help them steer products under development in a more usable direction.

The Book's Origins

The chapters of this book comprise an enhanced compendium of columns I have written as a contributing editor for *Medical Device & Diagnostic Industry* since 1991 and a few I wrote earlier for *Medical Design.* By 1994 my columns had covered a broad scope, and it seemed the right time to consolidate them into a book. I have updated these chapters as necessary to provide a current sense for user-interface design knowledge and practice as applied to the medical industry. I hope that this compendium provides design practitioners useful guidance on user-interface design processes and techniques and orients newcomers to the concept of user-centered design. I believe this is the first book of its type.

Scope and Organization

Considering user-centered design or the usability engineering process and associated techniques as a large puzzle, one might think of each of my chapters as the individual puzzle pieces. The book covers many of the key process issues, such as justifying the cost of usability work, and many of the important design techniques, such as designing effective graphical symbols. I have tried to keep my focus on medical design applications: products such as patient monitors, infusion pumps, cardiograph recorders, incubators, and digital thermometers. Recognizing the limits of a single book, I have included many references to other sources of design information and support. As such, I regard this book both as a primer on user-centered design and as a sourcebook.

The book is organized according to the following themes:

- Taking a user-centered approach to design
- The need for usable medical devices
- Determining user requirements
- Designing the user interface
- Prototyping the user interface
- Testing the user interface
- Documenting the user interface
- Special topics

The book first introduces user-centered design as a whole—providing the gestalt and then providing detailed guidance on design techniques and principles applied at various stages of the usability engineering process.

Those already indoctrinated in the usability engineering process may prefer to jump around to specific chapters of interest. Those who choose to read the book cover to cover will undoubtedly detect themes that repeat across several chapters, such as involving users early and continually throughout the design process.

Acknowledgments

This book has been made possible through the generous contribution of copyrighted material by Canon Communications, Inc., publisher of *Medical Device & Diagnostic Industry*, which has shown concern for the usability of medical devices for many years. I have several individuals to thank:

- John Bethune, Steve Halasay, and John Lehrer of *Medical Device & Diagnostic Industry* for supporting this book project and for their contributions to my earlier publications.

- My colleagues at American Institutes for Research (AIR), especially Joseph Dumas, vice president of AIR's New England Research Center, who have given me tremendous support during the years that I have been writing.

- Many other colleagues and fellow medical device developers, particularly Eric Smith, Datex Medical Instrumentation, and Larry Hoffman, Hewlett-Packard Company, who have contributed design examples, personal views, and unique insights to much of my writing.

- My wife, Pam, and my children, Ben, Ali, and Tom, for their love, patience, and support as I prepared this book.

Michael E. Wiklund
Lexington, Massachusetts
December 1994

Taking a User-Centered Approach to Design

Chapter

1

How to Implement Usability Engineering

The practice of usability engineering (or human factors engineering), which has been well established for about 40 years, seems to get "rediscovered" whenever an industry makes the transition from fundamentally electromechanical devices to those incorporating microprocessors in their designs. At the transition point, advances in microprocessor technology enable designers to extend the number and scope of device functions, resulting in more-complex products and a greater potential for usability problems and user complaints.

In order to anticipate possible usability problems, solve existing problems, and provide a generally higher level of customer satisfaction, companies turn to usability engineering, with its systematic research and design approaches, specific design techniques, and extensive database on human characteristics and performance. In general, as products become more complex, companies find that the need for usability engineering increases. The aviation, consumer electronics, and computer software industries have already adopted these principles; the transition is well under way in the medical device industry. Manufacturers that follow the procedures outlined in this text will discover that integrating usability engineering principles into the product development process is not difficult and frequently pays immediate dividends (see chapter 3).

Individuals or small committees that investigate usability engineering on behalf of their companies discover a variety of potential resources to assist them: numerous textbooks, undergraduate and graduate programs at universities, private consultants, and professional organizations (e.g., the Human Factors and Ergonomic Society, the Usability Professionals Association, and the Association for Computing Machinery/Special Interest Group on Computer and Human Interaction). Implementing usability engineering principles in an organization, it turns out, depends less on a company's business than on its corporate culture and level of commitment to its customers. The usability engineering process is fundamentally the same, whether one is developing aircraft instrumentation, software spreadsheets, or diagnostic scanners.

The Usability Engineering Process

The usability engineering process, in its myriad forms, builds upon traditional human factors engineering methods that some people associate with military systems development. An intrinsic element of usability engineering is an abiding respect for product users. To enable users to play a significant role in shaping product design, usability engineers become better acquainted with them by means of various research efforts—observations, contextual interviewing, group interviewing, and task analyses—that involve users. They complete the process by collecting feedback from people who use the finished product.

Ideally, usability engineering activities take place throughout the life cycle of a product development process. Practitioners have expounded several usability engineering models, most of which are similar (Nielsen 1993, 72). Here, the usability engineering process is presented in six steps:

- User studies
- Usability goal setting
- Concept development
- Detailed design
- Specification
- Field testing

Certain activities presented as stand-alone steps in other models, such as user-interface prototyping and usability testing, are incorporated in concept development and detailed design. Field testing is described here as a sixth step, although it can also be approached as a user-study activity—studying the ways that users interact with a finished product as a precursor to setting goals for a future product.

User Studies

Product developers are often tempted to jump directly into concept development, especially with fast-paced projects involving managers who are impatient to see tangible results. Even when they do pay attention to user research, product developers rarely place sufficient emphasis on usability issues, looking instead at matters that market research typically focuses on, such as feature set, pricing, and service. Often, the result is design shortcomings that create usability problems: moving a switch in the wrong direction, selecting the wrong menu choice, or not detecting an important change in a patient's condition as it is presented numerically or graphically on a vital-signs monitor.

Therefore, developers should resist the temptation to jump ahead with design, and instead conduct user studies first. Designers can use many different approaches to become acquainted with the user, some more exhaustive than others. Perhaps the simplest approach is just to watch people perform a sample of frequent, infrequent, urgent, and critical tasks with a medical device, and then talk to them about it. This process can yield important insights into device design. In some cases, designers might want to record in detail how users perform tasks using time-motion or other task-analysis techniques (see chapter 7). A follow-up interview with the same users in their work settings can provide additional information. Such contextual interviews can be conducted informally or in a more structured manner, using a prepared script in order to obtain the same information from a given number of users.

Conducting focus groups dedicated to usability issues, each with eight or so representative users, is another way for developers to get in touch with users quickly (see chapter 6). Working from a script of prepared questions, a trained moderator can lead users through discussions of important design issues and build toward a group consensus. Sometimes, designers use focus groups as a means of generating new product ideas. Obtaining one or two good ideas can make an .investment in focus groups worthwhile. The end product of a focus group is usually a report that serves as a body of evidence to support design decisions.

If designers need to collect a lot of detailed information about users and the tasks they perform with a product, a questionnaire might be the answer. For example, a patient-monitor manufacturer might want to find out how frequently users perform certain tasks, such as changing alarm limits, adjusting the time between noninvasive blood pressure measurements, or determining cardiac output. They also might want to ask users about their past experience with computers to determine how they might react to a computerlike user interface. The appeal of such surveys is that they produce a complete set of data that can be analyzed statistically. Their shortcoming is that they do not put developers in close communication with users, unless the surveys are conducted orally or followed by an interview.

The final products of a user study effort include such important but intangible results as a raised awareness among designers of users' needs and preferences, a stronger basis for making good design decisions on behalf of users, and greater empathy with people who use their companies' products. Perhaps the most important tangible product is a user profile—a document that describes users in a highly comprehensive manner. A user profile should answer a number of questions about users, including:

- Who are the users (age, gender, national origin, education, occupation)?
- What are their physiological characteristics?
- What are their psychological characteristics?
- What specialized knowledge and experience do they have?
- What jobs do they perform?
- What tools do they use to perform their jobs?
- How do they perform specific tasks?
- How do they interact with others?
- What is their work environment like?
- What is their level of motivation?
- What kind of experience do they have with similar products?
- What expectations do they have about learning to use a new product?

- What is their preferred learning style (Mayhew 1992, 30–79; Cushman and Rosenberg 1991, 18–20)?

User profiles should be shared among all design team members and reviewed frequently throughout the design process to ensure that evolving designs will be well suited to users (Figure 1.1).

Figure 1.1. *User interviews and testing revealed that a thumbwheel was the best way to achieve single-handed control of articulation for the illumination catheter of this endoscope.*

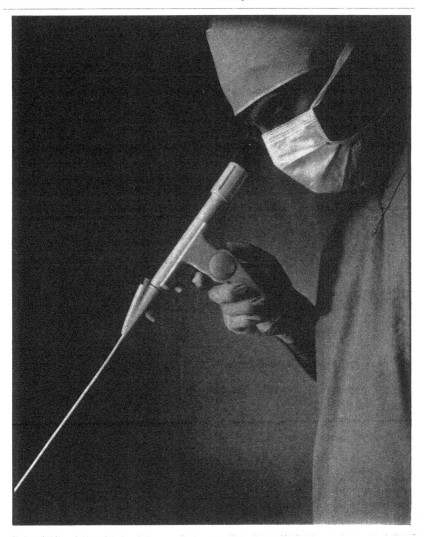

Courtesy of S. G. Hauser Associates, Inc.

Usability Goal Setting

Setting usability goals reflects developers' desires to follow a more quantified approach to usability, to be more concrete about what usability means and how to achieve it, and to ensure that usability goals are taken into account when design decisions and trade-offs are considered. Does setting usability goals accomplish these objectives? The jury is still out because of a shortage of experience applying the approach to real-world design projects. The concept of setting quantified usability goals has only gained momentum over the past five years or so. Today, however, many designers believe setting usability goals is a key to garnering credibility and resources in an R&D environment.

The notion is that designers can set goals for product usability in much the same way as they can set them for a product's size and weight. For example, just as designers might attempt to keep the weight of a portable patient monitor under 10 pounds or shoot for a minimum battery life of 3 hours of continuous monitor operation, they might attempt to limit the time required to perform a setup task to 30 seconds or less.

By setting goals before the design process commences, developers become more focused on developing a design that will achieve its goals. Assuming the goals will be reviewed and approved by design team management, an awareness of usability goals empowers designers to fight for the design resources and features necessary to achieve them. For example, a usability goal stipulating the speed at which a user will acquire information may dictate using a larger, costlier LCD display than originally planned—a trade-off favoring users.

Setting usability goals requires a period of adjustment for people unused to thinking about usability in quantifiable terms; so it is important to get these people involved in the goal-setting process. A good starting point is to list all of the major user tasks. Such a list can be drawn directly from the results of earlier task analyses or created from scratch, based on the product's proposed functions. Next, set goals for how long each task should take. Initially, these limits can be based on common sense. Preferably, however, they should be drawn from comparison or so-called benchmark tests of the company's current product or from a "best of breed" product from a competitor. Developers can set other types of task-specific goals, too—such as the successful rate of completion of a task or the number of errors committed while performing a task—depending on how a product is used and what aspect of task performance is most important.

At first, setting usability goals may seem to be based on guesswork. But the process becomes easier as developers accumulate experience with a class of products, such as patient monitors, and as they develop a larger base of usability test data. Sample goals might include the following:

- Eighty percent of users should be able to learn basic device tasks easily (that is, on a scale where 1 = poor and 5 = excellent, at least 80 percent of users should be able to give the device a rating of 4 or 5).

- On average, changing any alarm limit should take 15 seconds or less.

- Ninety percent of users should be able to prepare a device to perform an analysis without making an error.

Concept Development

Concept development is a crucial step in the usability engineering process. Ideally, developers begin with extensive knowledge about users and a set of usability goals, and end up with some promising design concepts. However, as with other fields of design (industrial design and architecture, for example), developers cannot simply plug data into a design algorithm and generate an effective solution. Quality design requires special training, talent, creativity, experience, and hard work.

That stated, a structured approach to concept development can facilitate good results. One possible approach is described below and includes the following five steps: explore mental models, develop a user-interface structure, add content, model the concepts, and evaluate the concepts.

Explore Mental Models

Good designs enable users to develop a simple and effective mental model (Figure 1.2) of how a product works. For example, to explain the topic of automobile power trains, someone not schooled in auto mechanics might say:

> When you start the engine and press the gas pedal, gasoline goes into the engine's cylinders. Spark plugs make the gas

Figure 1.2. *This drawing reflects a child's simple mental model of an automobile.*

Artist: Alison Wiklund, 6 years old

explode, moving the pistons. As the pistons go up and down, they make the crankshaft spin. The transmission takes the power from the spinning crankshaft and delivers it to the wheels.

The technical accuracy of such a mental model is secondary to its ability to help users understand and even anticipate how a product works. In the above example, for instance, it is not necessary for the listener to understand valve timing or the thermodynamics of combustion to gain a general understanding of automotive power trains. Similarly, it may not be necessary for a medical worker to understand the details of equipment calibration if calibration is an exclusively automatic device function or can be reduced to a minimum number of steps.

When designers speak of conceptual models, they are usually describing the underlying logic of an existing user interface or one in development. Conceptual models evolve from one or more mental models and form the fundamental basis for a design. Often, a conceptual model is built upon a metaphor, such as the filing system metaphor used by some software applications in which new electronic documents are "filed" in iconic folders analogously to the way real paper documents are filed. The metaphor enables users to map their real-world experience onto the use of a software product, thereby reducing the learning curve. Accordingly, user-interface designers should explore possible conceptual models and attempt to incorporate a unifying metaphor.

Appropriate conceptual models and metaphors may arise from user studies that ask representative users to explain how they regard an existing product and suggest analogs. Typically, when working on an all-new product, designers must develop a conceptual model based on their own mental models and then have users evaluate it.

Develop a User-Interface Structure

The objective of a user-interface structure is to illustrate the inter-relationships among the discrete functions of a device or software application. User-interface structures, for example, can be expressed in the form of bubble diagrams in which individual bubbles represent discrete displays (Figure 1.3). For software user interfaces, menu hierarchies and flowcharts can also be used. Ideally, the user-interface structure should be compatible with the conceptual model. That is, if the conceptual model divides a product's functions into three major groups, it may be appropriate to segregate associated controls into similar groupings, or to present three main menu choices on the top-level software display.

The design team's choice of user-device interaction style and pointing device can have a major effect on the user-interface structure of the final product. Designers should choose user-interface features based on how users will interact with a device, especially when designing a software user interface. For example, a software product's mechanisms for interaction may include selectable objects (e.g., icons) that enable direct manipulation, textual menus, command lines, or a question-and-answer dialogue. Also, designers must choose among alternative pointing devices, such as a touch screen, trackball, mouse, control wheel, keyboard, or even voice input.

The range of design options can make the concept development process, including developing an appropriate user-interface structure, seem unbounded. For example, the varied combination of 3 conceptual models, 3 user interaction styles, and 3 alternative pointing devices yields a possible 27 design concepts, which is too many to explore fully. One way to narrow the field is to ask key questions about important classes of variables during user studies. Depending on the product under

Figure 1.3. *Conceptual model reflected in a bubble diagram as well as the resulting layout of the product's control panel.*

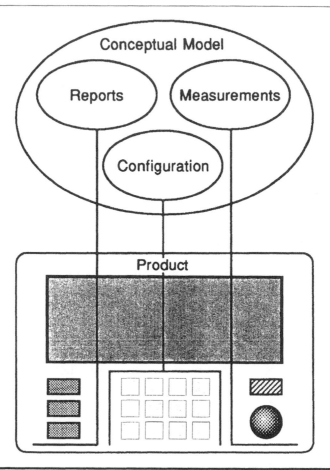

development and the design team's mission, 3 to 5 design concepts are probably a manageable number.

Add Content

A conceptual model and a user-interface structure form the core of a design concept. But it is hard to evaluate the merits of a design described only in terms of words or diagrams. Effective communication of hardware designs calls for drawings or physical models; software design concepts require screen designs, or even an interactive prototype (see chapters 21 and 22).

Accordingly, the core design must be enhanced with enough visual and functional details to give people a sense for how a product looks and works.

In terms of appearance and interaction style, software developers may want to follow guidelines such as those established for Windows-compatible products (*The Windows Interface* 1987). Hardware developers may want to match a new design with the appearance of an existing product line, or explore an entirely new look based on industrial design studies. When adding detail to a design concept, user-interface designers should use design principles presented in textbooks or guidelines documents, such as *Human Factors Engineering Guidelines and Preferred Practices for the Design of Medical Devices* (see chapter 10). They might also emulate examples of good user-interface design found in their own and related industries, assuming this can be done without violating patents and copyrights.

Model the Concepts

Industrial designers and architects use computer-aided design (CAD) software imaging and physical models to express their design concepts. User-interface designers work in a similar fashion, employing physical models and prototyping software such as Hypercard®, SuperCard™, Toolbook, Visual Basic™, Macro-Mind Director, and Altia® Design to communicate their designs. Building such models helps designers improve their designs; weak designs can be discarded in the process.

Because so much of a product's design can still change at this point, it is a waste of resources to build a fully functional prototype unless the product is quite simple. More often, the best course is to build a so-called low-fidelity prototype that gives people a basic sense of how the final product will work. Such prototypes can take a matter of hours or days to build. Prototypes of products incorporating a software user interface usually present users with a fairly complete top-level screen, and enable them to conduct a limited set of tasks.

If your company's development resources are limited and you intend to create more-robust prototypes, it makes sense to limit the number of concepts to about three. Users can focus their attention and make meaningful comparisons among three concepts. Also, whenever there are more than a few concepts, it is generally true that one or more are clearly inferior and could have been discarded earlier.

Evaluate the Concepts

The main purpose of modeling concepts is to obtain user feedback that can help the design team identify the most promising design direction. Model evaluation may identify one concept as the most promising, show that users prefer a hybrid of two or more concepts, or indicate the need to go back and explore other basic concepts.

Developers can productively evaluate design concepts by means of one-to-one interviews or focus groups with representative users; however, these can produce misleading results unless users have the opportunity to fully explore the design concepts in a hands-on manner. Therefore, a better approach is to conduct a usability test (see chapter 23).

The purpose of usability tests is to have users interact with concepts in order to measure user performance and preferences. The results of a usability test involving 5 to 10 representative users is usually definitive (Virzi 1990, 291–294). Of course, one can make a case for running a larger usability test. For example, one may want to increase the sample size if the differences among concepts are few in number or subtle, if members of the design team disagree strongly about which design is best, or if the design team wants to produce conclusions with a stronger foundation. In such cases 20 test participants seems to be the magic sample size that imbues credibility. Usability testing should, of course, be based on the usability goals set earlier. Subjects' performance times for given tasks should be compared to predetermined goals in order to evaluate how well the design concept meets the goals.

The test results can be conveyed to the design team orally at a meeting, but it is also a good idea to document the results in a written report. The report should couch design recommendations in the context of usability test results so that the recommendations do not seem arbitrary or biased. Reports can vary in length from a short memorandum to 50 pages, depending on the product, the audience, and the amount of detail provided. Test reports can serve as important tools for promoting a design direction that is in the best interest of users, as opposed to the best interest of an influential individual within the company.

Detailed Design

Let us assume that usability testing has identified the most promising concept among three prototypes. Turning the concept into a final design requires a lot of work, but is relatively straightforward. Again, designers should draw on established usability design principles, style guides, good examples found in the literature and at trade shows, and their design experience to produce a complete user-interface solution.

At this stage, however, designers need to be wary of losing their design vision, which is reflected in the prevailing conceptual model. For example, users may have preferred a design concept because it employed a simple card-file metaphor. But even simple metaphors can become complicated and corrupted when it comes time to implement all of the intended functions. Suddenly, the development team may end up with metaphors so extended or convoluted that they are barely recognizable— for instance, cards that have three sides, two scrolling windows, and hypertext links to information databases. These extended features may be quite useful considered individually, but they may interfere with users' ability to develop a simple mental model of how a product works. The goal is to start with a simple mental model, create a corresponding conceptual model, and then create a product that is faithful to the original mental model, so that users develop the same mental model the first time they use the product. Therefore, as product development progresses, designers should keep referring to the original conceptual model.

As is the case with concept development, detailed design development requires that the design team be involved with modeling and evaluation, but in a more rigorous manner. Affordable prototyping tools, such as those listed earlier, enable designers to produce realistic (i.e., high-fidelity) device prototypes, which in turn enable users to perform most tasks in a realistic manner. As a result, designers can construct usability tests (sometimes iterative) that are effective at uncovering usability problems, and that can still be corrected without undue penalty to a project's budget or schedule (Figure 1.4).

Specification

The final user-interface design must be documented in a way that enables designers, programmers, and engineers to implement it faithfully and other people to understand it. The most common options include reports, user-interface guidelines documents, layout drawings or templates, state diagrams, user-interface maps, screenplays, and prototypes.

Figure 1.4. *To refine this excimer laser system for interventional cardiology, by LAIS/Advanced Interventional Systems (Irvine, CA), product designers at S. G. Hauser Associates, Inc. (Calabasas, CA), conducted user testing that helped them to determine the best size for the horizontal desk surface, the size and location of graphics on the display screen, the positioning of sterile parts, and the location of the emergency on-off switch.*

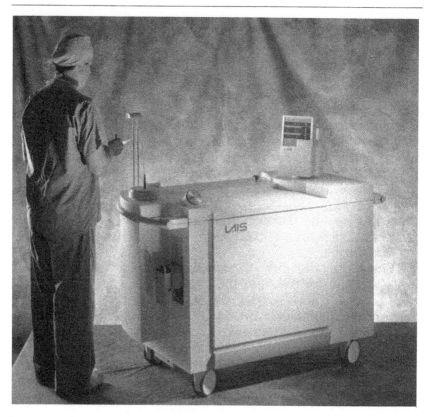

Courtesy of S. G. Hauser Associates, Inc.

A *report* explains the user-interface design, including the underlying conceptual model, the user-interface structure, the choice of interaction style and pointing device, and the detailed design features.

User-interface guidelines documents explain the rules for the way the user interface is structured. Such guidelines also help designers proofread the user interface for inconsistencies and provide a guide for implementing future design changes.

Layout drawings or *templates* show the placement of hardware and software features. Such drawings can be created by hand or by using computer-based drawing and CAD applications.

State diagrams present graphically the logical relationships between user actions and product responses. A state diagram helps designers proofread a user interface, making sure that all user inputs produce the expected responses and that there are no logical discontinuities.

The *user-interface map* delineates the way the user interface is organized, much the way organization diagrams reveal the way a company is structured. Such maps are particularly helpful in understanding the organization of menu-based software applications.

A *screenplay* is a set of screen prints generated from the prototype or produced using computer-based drawing packages. Screenplays are especially useful for tracking design changes once a prototype has been completed and is no longer subject to modification.

Prototypes are virtually complete simulations of the final product, modeling both appearance and interactivity with high fidelity. Such prototypes can usually be produced by modifying the prototype used for evaluating the detailed design. Any remaining differences between the specification prototype and the real product can be explained in a set of supplemental notes.

The choice of specification methods should be based on how formal your organization's specification process is and how useful the specifications are; these qualities can be gauged once you accumulate usability engineering experience. The breakthrough for many companies seems to be the point at which they are able to supplant endless amounts of text explaining how the user interface looks and works with a single, interactive prototype. After all, a textual description of a visual entity is subject to misinterpretation, while a picture (i.e., a prototype) is not.

Field Testing

Medical devices frequently go through a period of clinical testing before they are widely marketed. However, when a product enters clinical testing, it is usually much too late or too costly to make major changes to the user interface. Also, clinical testing does not usually produce systematic data about the performance of a user interface and users' impressions of it (see chapter 23).

Nonetheless, there is always the next version of a product to consider. Accordingly, testing the actual product can be thought of as either the end or beginning of the usability engineering process. Field evaluation or laboratory test data can help improve the realism of the usability goals for future versions of a product or related products.

Implementing a Program

Implementing a usability engineering program takes conviction. It is reassuring that several major corporations, particularly in the software industry, have committed themselves to this process (refer to Wiklund 1994, *Usability in Practice: How Companies Develop User-Friendly Products*, Boston: Academic Press). However, the transition can be difficult. Managers may not fully understand usability engineering and may place unreasonably high expectations on the outcome of the first trial project incorporating usability engineering principles. Then, unspectacular results may set the program back for years. Alternatively, companies may pay lip service to usability, then reject major innovations arising from the usability engineering process.

Experienced practitioners of usability engineering recommend picking the trial project carefully. The conservative approach is to choose a project that is not critical to the overall well-being of the organization, that involves managers who appreciate the importance of usability, that presents opportunities to produce a major payback (such as taking a product that has been heavily criticized by the media and customers as being hard to use and making it much better), and that can be accomplished on relatively short order and within a reasonable budget. If the project goes well—that is, if it produces useful results—then usability engineering will develop a reputation as being good for the organization, as well as for products and users.

A more aggressive approach is to apply usability engineering to a critical project. Often, the critical nature of the project is viewed as just cause to invest money in usability. Also, such projects are often initiated specifically because customers have complained that an existing product is hard to use and that a better product is urgently needed. Success on such a project can catapult a fledgling usability program into an established position.

Who should lead the trial project? Often, individuals from the fields of industrial design, software, or documentation step forward because of a personal interest. Such individuals frequently ask usability engineering consultants to help them plan and conduct the project.

Occasionally, companies simply hire a single usability specialist on an experimental basis. Such an individual usually reports to the R&D manager or the product manager, who may work in marketing. The specialist may also draw on consultant support, or take a so-called bootstrap approach, whereby he or she trains others within the company to help out with usability engineering activities. Rarely does a company start by forming a usability engineering group. That comes later, after the company is convinced the process is worth the investment.

Accordingly, publicizing the process and obtaining positive results are the keys to establishing a successful usability engineering program. People's natural interest in user-interface issues ensures an audience. Furthermore, if a company is applying usability engineering principles for the first time, the results are often dramatic. The lone specialist (who may also be a consultant) might start by teaching a usability engineering workshop for staff and management to increase understanding and receptivity. Next, he or she might invite managers to observe usability test results regarding an existing product to reinforce the need for improvement and build a database useful for setting subsequent goals. If a detailed design prototype is available, he or she might conduct a briefing to highlight how far the design has progressed and how the prototype provides a vehicle for communication and evaluation. If things go well, the product will be well received by the sales force and the end users and will generate brisk sales.

Once a development team has a successful track record applying a usability engineering approach, it can think about expanding its sphere of influence. Many managers within larger companies cannot imagine needing several usability engineering specialists, but they are frequently surprised to see how substantial a contribution a usability group makes, particularly if the company's products are complex or incorporate a software user interface.

Conclusion

Medical workers demand usability in the products they use professionally. For example, many nurses declare that if a product is not easy to use, they do not want it in their unit—they will find something else that works. In user studies nurses rate usability as one of their top design requirements (see chapter 4). This should send a strong message to all medical device manufacturers to invest heavily in usability.

Fortunately, implementing a usability engineering approach to design can be relatively straightforward. Trained professionals know how to implement the process in ways that do not detract from other elements of the development process, although the shift in "ownership" of the user interface may initially cause a few turf battles. The usability engineering process is fairly well defined, and the results can be measured, particularly if the design team sets usability goals and conducts usability tests to measure user-interface quality. It is no longer enough to state, "Our products are user friendly." For years, a majority of medical product manufacturers have been bleating this tired marketing claim in their literature and sales pitches with nothing to back it up. Today, providing credibility for such a claim depends on addressing usability issues in a realistic and comprehensive manner.

References

Cushman, W., and D. Rosenberg. 1991. *Human factors in product design.* New York: Elsevier Science Publishers B.V.

Human factors engineering guidelines and preferred practices for medical devices, AAMI HE-1988. 1988. Arlington, VA: Association for the Advancement of Medical Instrumentation.

Mayhew, D. 1992. *Principles and guidelines in software user interface design.* Englewood Cliffs, NJ: Prentice-Hall.

Nielsen, J. 1993. *Usability engineering.* Cambridge, MA: Academic Press.

Virzi, R. 1990. Streamlining the design process: Running fewer subjects. In *Proceedings of the Human Factors Society 34th Annual Meeting.* Santa Monica, CA: Human Factors Society.

The windows interface: An application design guide. 1987. Redmond, WA: Microsoft Press.

Wiklund, M. E. 1994. *Usability in practice: How companies develop user-friendly products.* Boston: Academic Press.

Chapter

2

Managing the Transition to a User-Centered Approach to Design

Setting up a usability engineering program represents a big step for any medical device manufacturer, regardless of its size or level of profitability. Taking this step acknowledges the importance of product usability in the overall scheme of total quality assurance and demonstrates the depth of a company's concern for its customers. It also can save time and money during product development and boost revenues later when improvements in usability generate increased sales.

However, in the short term, setting up a usability engineering program can really shake up an R&D organization. Shifting responsibility for user-interface development can bruise egos, because it suggests that the original individuals or departments responsible did not do a good job. Also, the time required to perform up-front usability engineering will test the mettle of most project managers, even those who believe in the value of a user-centered approach to design. Therefore, the transition to a user-centered design approach requires skillful and sensitive management by individuals who understand the politics of design. Depending on how the transition is handled, a usability engineering program either will be absorbed into the company's design culture and produce tangible benefits, or will flounder, causing product development delays and cost overruns, as well as dissension among design teams. This chapter will explain how to make the transition to user-centered design as smooth as possible.

Experimenting with Usability

To make sure their initial usability program efforts do not back-fire, managers starting a usability program should take advantage of the lessons learned by others. One way to do this is to talk to people who have pioneered usability programs in other companies, not necessarily in the medical industry. In fact, the current dearth of formal usability programs in the medical industry suggests that interested parties ought to look elsewhere—for example, to companies in the consumer software arena, where usability engineering programs are more common. Most of the major manufacturers in the consumer software industry (e.g., Borland, Lotus, Microsoft, and WordPerfect) have usability staffs. In general, usability pioneers are proud of their accomplishments and are pleased to share their experience with others, particularly if the people in need of advice are not direct competitors.

Most companies that have developed usability programs will state that full-blown programs are rarely created in one bold stroke. Such an approach goes against the conservative nature of many device manufacturers and requires a substantial investment in what is for them an unproven methodology. By contrast, most companies begin exploring the feasibility of usability engineering with a limited consultation or a small-scale demonstration project—a test drive of sorts.

For example, a project manager may have misgivings about a design in progress. Perhaps the prototype device's display has a particularly deep hierarchy, one that extends six or more levels, which causes users to become confused when they try to perform even common tasks. Or a design may have even more obvious usability problems, such as icons and labels that users misinterpret, or controls that require too precise a touch for most users to operate properly. These are typically the circumstances when project managers seek "professional help" from a usability specialist.

If a company decides to hire a usability specialist, he or she (or a consulting team) may spend from a day to a week reviewing a design in progress, applying an understanding of design principles and experience with similar products as a basis for suggesting design improvements. A consultation might begin with a meeting between the consultant and the project team (or just the project manager, if usability is already a contentious matter) to discuss the scope of required services and introduce the design problems. The consultant may be able to render an opinion about problems and potential solutions during the initial meeting or may need to perform additional analyses. Usually, subsequent to the meeting, the consultant will draw up a memorandum explaining the problem and potential solutions, augmented by design sketches or drawings, if appropriate.

At the next level of involvement, a consultant might lead a small demonstration project, in which 5 to 10 people objectively measure and compare the usability of a prototype device or several competitive products. In such instances it is advisable to test a product that has serious problems or that is crucial to the company's well-being in terms of sales volume or reputation for design excellence. That way, if the usability testing goes well, its value will be more evident than if an inconsequential product were chosen for improvement.

Even though medical devices are often used in specialized environments (e.g., the operating room, the intensive care unit, or the back of an ambulance), it is usually possible for researchers to simulate the necessary conditions and obtain useful results from testing in a controlled environment, such as a laboratory. For example, good data can be obtained in an office setting for devices as complex as anesthesia workstations. Alternatively, tests can be run in the field, as long as there is no health risk to operators or patients, or such testing does not interfere with important activities. If conducted with scientific rigor, a small project may cost $10,000 to $20,000. However, a less formal test involving a few subjects and limited data analysis and documentation might be accomplished for just a few thousand dollars.

Sometimes the first usability test turns out to be a watershed event. Project managers and design staff members alike are often amazed at what they learn from even a simple test, and their appetites are whetted for more. In order to increase the chances of positive outcomes, test planners should invite key managers to observe usability testing, or ask them to attend the presentation of results. Watching potential users struggle with a poor design and then succeed at using an improved one makes quick converts out of doubters. Because descriptions are no substitute for actually seeing people use products, usability testers often create a short videotape capturing the highlights of testing. Showing such tapes usually quickly settles any debate about the need for usability improvements.

If no funds are available for a meeting with an outside consultant, a less effective but still worthwhile approach is for a nonspecialist to evaluate the design according to the guidance provided in ANSI/AAMI HE48-1993, *Human Factors Engineering Guidelines and Preferred Practices for the Design of Medical Devices* (AAMI 1993), or an equivalent design guide. This is better than doing nothing. Because user-interface design is not a "cookbook" task, such an evaluation is likely to miss potential problems that occur when people use products to perform complicated tasks.

Evolving Toward a Permanent Program

Often, the next step companies take is to hire a permanent staff member or to retain a usability consultant on a long-term basis as an adjunct member of their design team for new product development efforts. Good arguments can be made for either approach, although hiring a new staff member is more likely to ensure long-term and consistent attention to usability matters. The work of consultants, no matter how effective, can be shut down if a company has a down turn in sales and the CEO dictates a corporate-wide freeze on elective spending. Full-time employees are more likely to survive cost-cutting measures.

A consultant's role might include conducting workshops for the staff on subjects such as the usability engineering process and applied design, generating user studies, leading the development of user-interface concepts, guiding the development of user-interface prototypes, and running usability tests. (See Table 2.1 for a list of tactical and strategic activities useful in starting a usability engineering program.)

Table 2.1. *Steps toward establishing a corporate usability program.*

Tactical

- Perform an expert review of a key product under development, resulting in a memo citing usability problems and potential solutions.

- Conduct a focus group to define user-interface requirements for a new product or to determine user preference among alternative user-interface design concepts.

- Conduct a usability test of an existing product, in order to benchmark performance; a computer-based prototype, in order to identify positive features and usability problems early in the design process; or a near-production working model, in order to validate its usability and resolve any residual usability problems.

Strategic

- Conduct a management-awareness briefing to explain the benefits and costs of a user-centered design process.

- Sponsor workshops on the usability engineering process, applied user-interface design, and testing in order to increase staff understanding and skills.

- Conduct an audit of the usability of all of the company's products to prioritize usability needs.

- Implement the usability engineering process on a key product development project as a basis for judging the benefits of such an approach.

- Hire a usability specialist to take control of the user-interface design process.

- Support the development of a more comprehensive usability program, perhaps including hiring a larger usability staff.

Holding workshops is a particularly good first step toward creating a permanent program, because it sends a signal to the staff that usability is important to management, it gets everyone oriented to and enthusiastic about the goals of the program, and it identifies ways for nonspecialists to play a role in the process. Sometimes it is best to conduct separate workshops—one for technical staff and another for those involved in management, marketing, and sales. That way, the focus of the workshops can be directed to the issues of greatest concern to each group.

When searching for a usability consultant, device companies should request that several experienced consulting firms provide them with detailed proposals that include a breakdown of project tasks, concrete goals, costs, and the identity of personnel responsible for specific activities. In order to facilitate comparison among firms, it is important to be sure that all consulting organizations bid on the same services.

A modest set of usability engineering activities could require an investment in the $30,000 to $100,000 range. Usually, consultants perform work on a time-and-materials basis or a cost-plus-fixed-fee basis. Taking a fixed-price approach can be problematic because usability programs often change direction in midstream as designers alter product requirements in response to user research findings. A compromise approach is to establish a time-and-materials contract with a ceiling amount not to be exceeded.

When usability consultants take the lead, members of the company's design staff often get heavily involved in assignments so they can learn the techniques and perform much of the work on subsequent projects. This bootstrapping or skill-transfer approach can work well, particularly if the design staff has the requisite talent and can increase its expertise by attending outside tutorials on usability (e.g., human factors engineering or experimental psychology). Some of the best tutorials are presented at the Association for Computing Machinery's Special Interest Group on Human and Computer Interaction Conference, which is held annually around April. The *Human Factors and Ergonomics Society* (HFES) *Bulletin,* a monthly newsletter, often lists or advertises short courses beneficial to usability professionals who want to expand their skills.

Hiring Full-Time Staff

If a company chooses to hire a full-time specialist, the new hire might begin by performing roughly the same activities as an outside consultant. However, he or she would do well to pay close attention to building long-term working relationships with the design staff and, in the process, becoming as visible in the company as possible. Usability specialists need to be visible because project teams are rarely required to use the services of usability specialists. Rather, usability engineering is presented as a corporate resource, much the way industrial design is positioned in many companies, and project managers can choose to use the service or not. Accordingly, usability specialists must continually sell their services on the basis that they add value to products.

One way usability specialists can promote themselves is to direct their efforts to the top management of an organization. If the company CEO or other high-level executives take a personal interest in usability and voice their strong support, these executives can provide the program with a tremendous shot in the arm, which can result in project leaders setting aside money in their development budgets for usability engineering services. A second way for usability specialists to achieve visibility is to conduct a continuing series of lunchtime lectures on usability, inviting guest speakers, whenever possible, to bring in a fresh perspective. A third way is to locate the usability testing facility (even if it is just a single room) in a central location and welcome people to drop in whenever they like, whether or not they are involved with the product being tested at the time. A fourth way is for the usability specialist to plan individual meetings with project managers and make a sales pitch for usability services.

If a majority of project managers do not feel that usability engineering is making a meaningful contribution relative to its cost, the program risks being terminated. Therefore, companies should hire specialists with extensive consulting experience, people who can fulfill raised expectation with successful design solutions.

Usability specialists should document and then showcase the benefits of their work. For example, they should prepare "before and after" examples of product designs and, with hard data from performance testing, reinforce the intuitive sense of others in the company that usability engineering has produced design improvements.

Also, specialists should carefully track how much usability engineering actually costs, and then project cost savings and added revenues to establish a cost/benefit ratio. Although few companies require that usability engineering groups justify the costs of their existence, circumstances can change quickly. Companies that have a bad year or even a bad quarter may suddenly require all departments to justify their existence.

Finally, usability specialists should establish good client relations, including a mechanism for feedback with their internal customers. This way, the specialist will be the first to know if a client is unhappy, and will be able to do something about it.

How much does it cost to bring an experienced usability professional on board? According to surveys conducted by HFES and recent job advertisements, senior usability specialists with at least 10 years of experience earn from $60,000 to $100,000 or more per year, depending on the area of the country, the type of work, and the extent of their management responsibilities (Sanders 1993, 1–3). By comparison, entry-level people with an M.S. or Ph.D. in a usability field but no work experience typically start at an annual salary between $30,000 and $45,000.

Establishing a Solid Base for Usability

In order to empower usability specialists, it makes sense for management to bring them into the organization at a level equivalent, say, to the manager of industrial design or technical communications. Bringing someone in at a lower level tends to subordinate the value of the usability program and hinder the specialist's ability to battle for the benefits of usability engineering should resources become limited.

In addition, bringing in usability specialists at a low level complicates matters if the usability group expands. For example, a usability contingent started in an industrial design group may grow to be as large as or larger than the host group. This might lead to battles for a share of resources, which could be avoided if the groups were originally established as organizations of equal importance.

Companies with an informal culture tend to position usability managers higher in the organization than those with a more formal, hierarchical culture. Perhaps this is because top-level executives in informal organizations are more likely to be involved in day-to-day design activities, or at least make themselves accessible to lower-level managers looking for support for their usability initiatives. Given such involvement or access, if a top executive is excited by the potential of a usability program, he or she may shepherd the concept into reality, in part by giving the usability manager a higher-level job. This, in turn, increases the chance that the usability program can compete successfully for resources and contribute effectively to product development efforts.

If one or more demonstration projects go well over the course of six months or so, they add impetus for expanding the role of usability specialists, particularly if the costs and benefits are well documented. Some companies may continue to use external consultants, especially if project managers have developed confidence in their abilities.

Companies that choose to develop an internal group of usability specialists tend to grow at a pace of one or two new positions a year. Often, the second or third person to join the usability team assumes total responsibility for usability testing, so group leaders may want to find an individual strongly interested in research and evaluation, as opposed to design. Additional group members may be hired from the outside or may transfer from internal industrial design or technical writing groups.

In addition, many companies reach out to local universities for student interns seeking hands-on experience in exchange for a modest wage or course credit if the work is research related. A list of universities offering usability-related programs can be obtained from HFES, as can a list of HFES student chapters. An HFES newsletter announcement of an internship opportunity usually garners a strong response (Table 2.2).

As a usability group grows, skill mix becomes an issue. Some groups are homogeneous, composed exclusively of usability specialists with backgrounds in human factors engineering,

---

Table 2.2. *Sources of usability engineering services.*

What to do when you need to find a usability engineering specialist:

- Make contacts at the annual meetings of organizations such as the Human Factors and Ergonomics Society (Santa Monica, CA, 310-394-1811), the Usability Professionals Organization (Derek Hoiem, Microsoft Corp., Redmond, WA, 206-882-8080), or the Association for Computing Machinery's Special Interest Group on Human and Computer Interaction (New York City, 212-869-7440). Also, these organizations may have lists of individuals and firms that provide usability engineering consulting services.

- Attend medical standards meetings that focus on usability issues, such as the development of harmonious alarms for medical devices or design conventions for anesthesia workstations. These meetings are sponsored by organizations such as the American National Standards Institute (New York City, 212-642-4900) and the Association for the Advancement of Medical Instrumentation (Arlington, VA, 703-525-4890).

experimental psychology, and the like. Other groups include a core of usability specialists, complemented by persons who have writing, graphic design, or user-interface programming skills.

As the size of a usability staff increases, companies may need to decide whether to maintain a centralized usability department or distribute usability personnel among project teams. Most companies maintain a centralized group, usually because the demand for usability services tends to outstrip the supply if things are going well for the company and there are new products in the pipeline. In such situations it does not make sense for a specialist to work exclusively for one design team, because other teams may not obtain the service they require. Furthermore, centralization facilitates professional development of the staff, particularly of less-experienced members of the group. Also, a centralized group is better able to address non-project needs, such as developing corporate standards for user-interface design or representing the company on industry groups working on user-interface standards and guidance materials, such as the American Society for Testing and Materials Committee F29 on Anesthesia and Respiratory Equipment.

If it is not feasible to develop a large usability group, one to three core specialists often manage a network of consulting support. This approach has the advantage of enabling smaller companies to vary the scale of usability support to the company's current needs, tap into the necessary expertise, and maintain continuity.

A Manufacturer at the Threshold

Today, North American Dräger (Telford, PA), a manufacturer of anesthesia products, stands at the threshold of creating a structured usability program. Until now, their approach to usability has been less formal, relying on the talents of marketing and engineering staff members, individuals who have no formal usability training but who have experience working with customers to define their needs. The company has also drawn usability support from industrial design firms retained to address product packaging and aesthetic issues.

According to Abe Abramovich, vice president of engineering, this approach has worked well for the company.

> We've been very successful bringing products to market that have met users' needs, although not every product has been a

home run from a usability standpoint. Some have been triples. However, we recognized that we could do an even better job—particularly at making our user interfaces consistent—if we hired a specialist to captain the effort and chart a course for the future.

Abramovich says that the increasing sophistication of anesthesia products incorporating computer displays was the key factor in the company's decision to hire an in-house specialist. Also, the company is paying greater attention to customer feedback. Customers report that they need the product information on the company's anesthesia workstation to be well organized and easy to understand so they can make sense of it quickly and effortlessly during critical periods, when stress is high.

The specialist hired by North American Dräger will enter the organization on a level with program managers and will report directly to Abramovich.

> By bringing the usability specialist in at a fairly high level, we will give that individual the necessary power and authority to be effective.

The specialist's agenda will include both tactical and strategic activities. Tactically, the specialist will need to contribute to design projects in progress, resolving design issues such as the wording of a warning message, the selection of a pointing device, or the layout of a remote display. Strategically, the specialist will need to define his or her long-term role in the company, which could include building a more robust usability program requiring additional staff. Abramovich expects the specialist's work to cut across engineering and marketing lines and, therefore, to require strong communication skills.

Abramovich admits that finding the right person for the job has been a challenge.

> I have received a mountain of resumes. Many of the candidates have an extensive publication list related to their academic work, but most lack product development experience. We need someone who has had hands-on experience developing several products, someone who can immediately help our development team make good decisions.

Accordingly, Abramovich has expanded his search to include usability professionals who may not necessarily have a background in medical product design but, nonetheless, have a product design-oriented background.

Conclusion

If the trend toward total quality management continues, usability will emerge as one of the driving factors in the design process, if not the hub of design activity. This is because usability concerns are so central to meeting customers' needs. The emergence of usability programs in response to quality goals is already occurring within some of the more progressive companies that build complex devices. Their emphasis on usability and the resulting product improvements will raise medical device users' expectations for usability, placing pressure on all manufacturers to deliver a high level of usability in their designs. Accordingly, the time is ripe for companies to plan their usability strategy—to determine how they are going to make the transition to a more user-centered design process.

In other industries, companies starting usability programs have taken a series of small steps, ultimately leading to the establishment of a substantial program. Early on, they have drawn effectively on consulting support, then hired a seasoned usability specialist to take full responsibility for usability leadership and expand the scope of usability services delivered to product development teams.

Usually, this means building an in-house group of specialists who may also draw on outside consulting support to serve the organization's needs. The vitality and contributions of such groups depend heavily on energetic leadership, a receptive corporate environment, and taking initiatives that are likely to produce the biggest payoffs. Most groups establish their worth quickly from both a financial and design-quality standpoint. Once such programs are established, companies wonder how they ever got along without them.

References

Human Factors Engineering Guidelines and Preferred Practices for the Design of Medical Devices, ANSI/AAMI HE48-1993. 1993. Arlington, VA: Association for the Advancement of Medical Instrumentation.

Sanders, M. 1993. 1993 Salary Survey. *Human Factors and Ergonomics Society Bulletin* 13(11):1–3.

Chapter

3

Measuring the Usability Payoff

For many medical device manufacturers, controlling product development and support costs is tantamount to survival. This is particularly true for companies serving mature, competitive markets, in which most products perform about equally well. In such markets overpriced products—which can result from a protracted development effort—may languish. To stay competitive, therefore, many medical device manufacturers are driven by the pressure to keep prices down and make a profit.

A product development culture driven by the bottom line is a difficult environment in which to promote a new usability engineering program, unless the promoter can quantify the program's costs and benefits. In these circumstances usability engineering specialists have learned to sell their programs not simply on the principle that usability is a good thing, but also on the basis of sound economic considerations.

Recently, Deborah Mayhew, a consultant based in West Tisbury, MA, has helped usability specialists justify the cost of their services. She has coedited *Cost Justifying Usability* (Bias and Mayhew 1994). Mayhew suggests that usability specialists make conservative estimates of usability benefits and then work backward to determine the appropriate amount of spending required to achieve them. Thus, she uses cost justification as a way of planning how much can be safely invested in a usability program. She then suggests translating the estimated benefits into concrete, quantified usability goals. For example,

> Determine the cost savings resulting from a user interface design that requires 30% less customer training time. Presuming you set conservative design goals, you can use the analysis to decide how much or how little to invest in usability testing.

> Another benefit of making conservative benefit estimates is that you're more likely to convince your audience of the value of usability testing. If you claim large expected benefits, your audience might reject your whole analysis. If, however, you assume modest benefits but still show a payoff, your audience is more likely to accept your analysis and grant the resources you're trying to obtain.

Joseph Dumas and Janice Redish (American Institutes for Research, Lexington, MA) also address the cost justification issue in *A Practical Guide to Usability Testing* (Dumas and Redish 1994) They view usability testing as a key step toward quantifying usability benefits and advise building a business case to convince managers that usability engineering is worth the expense.

> Usability engineering does have costs associated with it. However, don't just count the extra costs on the development side. Balance new costs with potential savings both before and after the product is released.

In order to build a business case for usability engineering, usability specialists must project the potential benefits of such a program for the devices their companies manufacture. Then they can measure the costs of implementing those benefits against the savings the benefits are likely to produce.

Potential Benefits

The potential benefits of usability engineering include reduced development time, better user documentation, improved marketing and sales, reduced training requirements, fewer product returns, reduced customer service requirements, extended market life, and liability protection.

Reduced Development Time

For software products, such as a modern spreadsheet or word-processing application or even an anesthesia record-keeper, the user interface may account for 60 percent of the software code (Karat 1992, 2). Therefore, major design changes that occur after preliminary coding can make a lot of extra work and delay a product's release. For example, the need to eliminate inconsistencies in the syntax of user inputs or the reformatting of on-screen information may add weeks to the development schedule. A more sweeping change, such as reorganizing a menu hierarchy to be more task oriented, may add months. Such delays should be a matter of great concern to manufacturers, considering that

> companies generally lose 33% of after-tax profit when they ship products six months late, compared with losses of 3.5% when they exceed development budgets by 50% (Karat 1992, 4).

The costs of increased development time should be measured in terms of both internal costs and missed sales. Accordingly, developers should be sure to integrate usability engineering into the early stages of a product development process.

Better User Documentation

Most writers will tell you that a good user-interface design helps them to write better documentation in less time. If the user interface is well designed, writers do not have to waste time figuring out how to explain complex and awkward device functions. As a result, they may be able to write shorter user documentation and production, packaging, and distribution costs may be reduced as well.

Improved Marketing and Sales

Usability engineering can benefit marketing and sales in several interrelated ways. For example, many medical devices are marketed at trade shows and technical conferences. In such circumstances where potential customers can see, touch, and perhaps try out products, if a device is easy to use, it may increase customer enthusiasm, facilitating a sale. Medical devices are also loaned to institutions or individuals for trial-use periods, during which potential customers are able to try out a product's user interface without the expert assistance of a salesperson. In some cases an institution may conduct systematic product comparisons as a basis for making a final purchase decision. For example, hospital personnel may evaluate several different patient monitors in their operating rooms, using each one for a few weeks, then rating its performance. Good product performance under both circumstances is sure to benefit sales.

Finally, in certain industry sectors, companies commission usability tests exclusively for the purpose of generating marketing claims to facilitate sales. When usability engineering is part of the normal development process, such claims can be developed at little added cost. Also, the mere existence of a usability engineering program can contribute to customer goodwill, since it reflects concern for customers' needs.

Reduced Training Requirements

Manufacturers spend a lot of money in training people to use their products. At the time a product is introduced to the market, a manufacturer may provide an initial series of seminars or workshops for clinicians, followed by repeated in-service training sessions. A better user interface might not eliminate the need for this training, since they suit the learning style of many users (see chapter 4) and have ancillary marketing benefits. However, an intuitive design may reduce the length of the training sessions while increasing their effectiveness.

Fewer Product Returns

Sometimes, end users who are unable to get a product to work properly conclude that it is broken. They may not even take the time to read the user manual to solve the problem. They may return the product to the manufacturer for a refund and adopt a negative view of the manufacturer's entire product line. Such an outcome can be costly to a manufacturer and damaging to its reputation.

Reduced Customer Service Requirements

Some companies operate toll-free customer hotlines and provide on-site service to help users solve equipment problems. Particularly virulent usability problems can lead many customers to call a manufacturer with the same complaint. Eliminating usability problems before a product is brought to market will reduce the likelihood of such complaints, and reduce labor and related costs such as phone charges and travel expenses for service representatives.

Extended Market Life

Frequently, manufacturers are pressured to introduce products that have extensive usability problems (known and unknown). Invariably, initial customers who encounter these problems complain to the manufacturer and may have a lower opinion of the product. They resent serving the de facto role of usability test subjects. In such circumstances manufacturers may be forced to update a product earlier than had been planned and provide customers with free upgrades.

Liability Protection

Data from the government's medical device reporting system show that a large percentage of reported incidents involve some form of human error (see chapter 5) that can be traced to deficiencies in user-interface design. Such deficiencies present a cost in the form of pain and suffering to the victims of accidents and a financial cost to manufacturers confronted with liability claims.

Whether liability claims are upheld in court may hinge on whether a manufacturer is able to demonstrate that it has taken reasonable care in the design process. In a deposition or court appearance, design managers may be asked: Did you conduct any usability engineering studies aimed at identifying potential human errors, then develop the appropriate safeguards? An affirmative answer can be quite helpful in a manufacturer's defense.

Moreover, usability testing and resultant product modifications might have prevented the accident altogether and averted the liability claim. Quantifying the potential benefits of such protection must take into account the number of products a manufacturer produces, the chance of a claim being filed, and the potential settlement size. Clearly, averting a single settlement of over $1 million would justify a substantial usability engineering program.

Cost/Benefit Analysis

Four steps in analyzing the likely costs and payback of achieving these kinds of usability improvements are as follows:

determining the cost of usability engineering, measuring usability improvements, estimating the value of such improvements, and calculating the return on investment.

First, calculate the cost of any usability engineering performed during the product development effort. For example, the internal costs of a comprehensive usability engineering program supporting a major product development effort may be about $130,000 (Mantei and Teorey 1988, 431). Approaching usability engineering from a task-by-task perspective, a company might spend $5000 to $10,000 on a focus group or design review, $20,000 to $50,000 to design and prototype a user interface, or $10,000 to $25,000 for a comprehensive usability test of a medical device of moderate to high complexity. When developing cost estimates for internal work, remember to calculate in terms of burdened labor costs and to include other operating expenses, such as the amortized cost of constructing and operating a usability test laboratory (see chapter 24).

Next, determine the effect of usability engineering on the usability of the final product by measuring user-interface performance before and after usability engineering input. Dumas and Redish suggest conducting a usability test on successive iterations of a product in development (Dumas and Redish 1994). For example, after several iterations, refinements to a blood gas analyzer's user interface may produce a 20 percent improvement in the time required to program a blood test, the rate of user error on a critical task, or ratings of the product's overall ease of use.

Usability experts must next translate usability improvements into manufacturer benefits. For example, one might ask: Will a 20 percent decrease in the time required to program a blood test increase overall user preference and improve sales? Intuition suggests that it may and product-specific market research data may confirm it. For example, a customer survey may find that ease of use is a chief factor in purchase decisions. Accordingly, a 20 percent decrease in the time required to program a blood test may result in a 10 percent increase in sales. As Mayhew suggests, this projected increase in sales could become the goal of a usability engineering program, with the additional profits and target benefit-to-cost ratio forming the basis for a program budget.

Of course, the benefits related to improved usability are likely to extend beyond increased sales to include, for example, reductions in training time and necessary customer support.

Therefore, the full range of possible benefits must be evaluated and quantified before completing program budgets.

Finally, develop a benefit-to-cost ratio. For example, one might tally profits from increased sales and reduced customer support costs over the first year of product life and compare it to internal costs of the usability engineering program. In accordance with a company's commitment to product quality, a breakeven outcome well into the market life of a product would seem sufficient to justify the costs of usability engineering. Figure 3.1 shows the breakeven point occurring much sooner (shortly after product release) for a hypothetical product development effort.

Recently published results suggest there may be a substantial payoff for usability testing within a year of product release. In one study an IBM researcher analyzed the benefits and costs

Figure 3.1. *Cumulative savings generated by a usability engineering program. The downward slope (a) reflects the cost of usability engineering, the upward slope (b) reflects savings accruing from a reduced number of late design changes, and the sustained increase (c) reflects savings from recurring benefits (e.g., reduced training and customer service costs).*

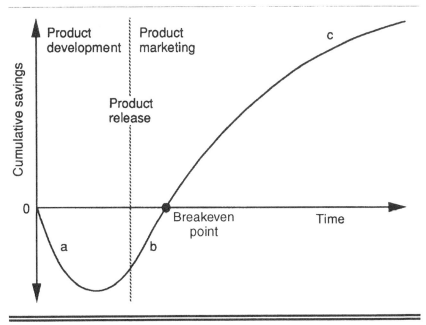

of improving the usability of two software applications developed for IBM's internal use (Karat 1990, 839–843). Over the course of one development effort, a tight schedule led usability engineering specialists to apply low-technology human factors techniques, including repeated usability testing of so-called foil prototypes (static overhead projections of user-interface designs), to evaluate and refine a software application used by marketers. The effort cost $20,700. Based on a 4.67-minute reduction in the time it took users to complete a task, task frequency estimates, an end-user population of 22,876 people, and personnel cost data, IBM calculated a savings of $41,700, or a 2:1 benefit-to-cost ratio, after the first year of application use.

Over the course of a second development effort, usability engineering specialists took more time and applied higher-technology human factors techniques, including more-exhaustive usability testing employing an on-line prototype to evaluate and refine the user interface of another software application. The second effort cost $68,000. Reductions in task performance times were projected to save $6.8 million in the first year, yielding a 100:1 benefit-to-cost ratio.

While device manufacturers may not accrue direct benefits from improved task performance, their claims of performance improvement can have a strong influence on customers' purchasing decisions, particularly when they can be substantiated by data from usability testing.

Conclusion

Some manufacturers view usability engineering as an expensive and dispensable luxury, even though they probably have not closely examined its benefits and costs. When R&D budgets are tight, manufacturers often fall back on commonsense design practices that frequently fail to produce a usable product. As a result, companies may be confronted with late design changes that delay the release of a product; an increase in customer complaints and support requirements; and pressure to replace an existing, flawed product with an improved version.

Other manufacturers view usability engineering as an essential and wise investment, based on conservative benefit-and-cost analysis. When R&D budgets are tight, they still spend the time and money necessary to develop a high-quality user interface on the first try. Their conservative estimates show that an up-front investment is likely to pay off several times over. When their products are released, they expect to spend less on training, customer support, and near-term upgrades because their usability testing shows that the product meets the users' needs. Also, they expect that their products will fare well and perhaps have a competitive edge over those of other device manufacturers.

Of course, not every usability engineering effort is going to produce a return on investment. Some efforts may reflect design overkill or be poorly executed. However, according to recent studies, well-planned and properly scaled usability engineering efforts are likely to be money makers. Manufacturers should, therefore, feel encouraged to develop conservative estimates of the benefits and costs of a usability engineering program.

References

Bias, R., and D. Mayhew, eds. 1994. *Cost justifying usability.* Cambridge, MA: Academic Press.

Dumas, J., and J. Redish. 1994. *A practical guide to usability testing.* Norwood, NJ: Ablex Publishing.

Karat, C. 1990. Cost-benefit analysis of usability engineering techniques. In *Proceedings of the Human Factors Society 34th Annual Meeting.* Santa Monica, CA: Human Factors Society.

Karat, C. 1992. Cost justifying human factors support on software development projects. *Human Factors Society Bulletin* 35(11):2, 4.

Mantei, M., and T. Teorey. 1988. Cost/benefit analysis for incorporating human factors in the software life cycle. *Communications of the ACM* 31:431.

The Need for Usable Medical Devices

Chapter

4

Medical Device Usability: A Survey of Critical-Care Nurses

Medical devices perform remarkable, life-sustaining functions in critical-care environments, such as an intensive-care unit. More remarkable than the capabilities of these medical devices, however, are the capabilities of the critical-care nurses who use them. Critical-care nurses perform their jobs under intense pressure that comes from working with very sick patients and from the need to set up and manage the functions of a large number of medical devices at the same time. A single device, such as a physiological monitor, may require that a nurse make numerous connections between the monitor and patient as well as adjustments to a slew of controls and displays. Furthermore, a postoperative patient may at one point be simultaneously connected to several complicated devices (e.g., a respirator, a physiological monitor, an intraaortic balloon pump, and several IV pumps). Therefore, the critical-care nurse needs to be a skilled technician as well as a capable clinician.

Because the high exposure level of critical-care nurses to medical technology gives them a strong, firsthand basis for assessing its usability, progressive manufacturers frequently seek usability assessments of their medical devices from critical-care nurses, usually in the form of clinical trials. Unfortunately, findings from clinical trials are usually limited to the users' views of a specific product and are treated by manufacturers as proprietary, and nurses' views on the usability of medical devices in general are not well known. To address this issue, I conducted a survey of critical-care nurses on a variety of usability topics, ranging from the impact of technology on nursing practice to the proper balance of device functionality versus operational complexity.

Survey Particulars

The survey was conducted among nurses in the following Boston-area hospitals in March 1992: Symmes Hospital, Arlington (110 beds); Emerson Hospital, Concord (188 beds); Newton Wellesley Hospital, Newton (379 beds); New England Deaconess Hospital, Boston (461 beds); New England Medical Center Hospital, Boston (474 beds); and Brigham & Women's Hospital, Boston (726 beds).

The survey form included 16 multiple-choice and fill-in-the-blank questions relating to the usability of medical devices. A cover page instructed nurses that their identities would be protected and that the survey results would be reported initially in *Medical Device & Diagnostic Industry.* Survey forms were delivered to interested nurse managers, who distributed them on a voluntary basis to members of their critical-care nursing staffs.

Thirty-seven nurses from intensive-care, cardiac-care, and surgical intensive-care units responded, which represents a small percentage of the total number of critical-care nurses working at these six hospitals. Furthermore, many of the respondents from each hospital were from the same nursing shift and critical-care unit. The ages of the nurses ranged from 24 to 54 years, averaging 35.3 years. Nursing experience ranged from 1 to 20 years, averaging 8.7 years.

Results

Manufacturers' Emphasis on Usability

A majority (60 percent) of critical-care nurses credit the medical device industry as a whole with paying close attention to usability concerns. Nonetheless, from their responses it seems that much more could be accomplished. A significant minority (40 percent) feel that product usability can be further improved through greater manufacturer commitment.

To their credit a majority of nurses expressed an interest in working with manufacturers toward improving medical device usability. When asked how they would respond if a manufacturer invited them to contribute ideas to the design of new medical devices (e.g., through an interview, a focus group, or a product test), 59 percent said they would be "very interested" and 38 percent said they would be "somewhat interested" in contributing. This finding suggests that manufacturers could benefit from creating additional opportunities for nurses to participate in the design process beyond the traditional clinical trial, which comes too late in the product development process for nurses to have a significant effect on the quality of user interfaces. For example, manufacturers can involve nurses in the design process via focus groups, contextual interviews, and usability tests of designs under development. Manufacturers might also retain nurses as clinical specialists or to serve on so-called advisory panels.

Effects of Technology on Nursing Workload and Practice

New medical technology is often introduced with the promise of labor savings, which presumably would allow nurses more time with their patients. When asked how recent advances in medical device technology have affected their overall workload, 56 percent of the nurses said that technical advances have decreased their workload, 30 percent said that advances have had no effect on their workload, and 14 percent said that advances have increased their workload.

Thirty percent of the nurses interviewed felt that technical advances have increased the amount of hands-on care they can provide to patients, 49 percent felt that advances have made no difference in this matter, and 21 percent felt that technical advances have decreased the amount of direct care they can give to patients.

These findings suggest that the labor savings provided by new technologies are being diverted from direct patient care, perhaps toward ancillary tasks such as medical record-keeping.

Preferred Learning Strategies

The survey indicated that the nurses have a strong preference for learning strategies that involve personal interaction with another human being. The two most popular approaches to learning device use were through an in-service demonstration or help from a coworker; learning on one's own (with or without assistance from a user manual) was a much less popular choice. This preference is likely to be at odds with the strategies of device manufacturers who are striving, as a way of lowering expenses, to reduce the amount of on-site customer training they provide. This fact should also give pause to manufacturers of computer-based products who plan to replace on-site training with an on-line help system. As one respondent noted on her survey form:

> It is extremely important for a nurse to be able to ask questions.

These findings indicate how different the nurses who use medical devices are from the engineers who design them. In general, design engineers learn how new products work by exploring the user interface or, as a last resort, by asking a colleague for help when they get stuck. They resist having someone show them, in a step-by-step fashion, how to use a new product, regardless of whether the demonstrator is a product expert or a coworker. The learning style that many engineers prefer most, namely, using a device by themselves without a manual, is the style that is least preferred by nurses.

Not only do nurses want person-to-person training, they want more of it. When asked how they would adjust the amount of training they receive on the use of medical devices, 54 percent said they would increase the amount of training, 46 percent said they would maintain the same amount of training, and none said they would decrease the amount of training they received.

Proficiency Using Medical Devices

Nurses tend to be "can do" people who bring self-assurance to their work; something most patients are liable to appreciate. Unfortunately, this makes it harder for them to admit shortcomings in their ability to use a medical device or to criticize a device as being hard to use. This reluctance is one reason that manufacturers receive positive feedback (or minimal negative feedback) from nurses on user-interface designs that human factors experts consider seriously flawed. For example, nurses who struggle to learn to use a particularly perplexing device still might rate it as reasonably easy to learn to use so as not to seem critical or incompetent.

Therefore, it is not surprising that the nurses surveyed reported a high degree of mastery over the medical devices they use on a daily basis. On the average the nurses felt that they operate 92 percent of the medical devices they use everyday with sufficient proficiency that they would feel comfortable teaching someone else how to operate them.

However, even assuming this perceived level of proficiency to be reflective of actual performance, one must ask whether this degree of mastery is high enough. After all, human error in the use of a medical device can have serious and perhaps irreversible consequences. Therefore, if nurses sense gaps in their abilities to use 1 in 10 medical devices, it stands to reason that more, not less, training is in order to reduce the potential for error (see chapter 5).

The potential for error, precipitated by insufficient mastery of device functions, appears to increase with device complexity. On the average nurses felt comfortable teaching someone else how to operate 85 percent of the more-complex medical devices they use with proficiency on a daily basis.

The drop in proficiency associated with more-complex devices may be attributed, in part, to the fact that an increasing number of medical devices are actually special-purpose computers. As such, these advanced devices function in a totally different manner from conventional, hardware-based devices. Special-purpose computers, such as patient monitors, require users to develop a mental model of how functions are organized within layers of software. Because much of the functionality is hidden, users rarely discover the full power of a given device. This leaves them with feelings of uncertainty and a reduced sense of control of device functions.

It is not surprising, then, that fewer than half of all respondents (41 percent) feel "very comfortable" using computer-based medical devices. Of the remaining sample, 35 percent said they were "comfortable," 16 percent were "neither comfortable nor uncomfortable," 8 percent were "somewhat uncomfortable," and none reported being "very uncomfortable" with such devices. Improvements in nurses' confidence and abilities to use computer-based medical devices may come with increased exposure to such devices.

Formal training is also important. A significant percentage of nurses (44 percent) reported that they had not received sufficient formal training in the use of computer-based devices. Apparently, such training does not yet occur to a great extent in nursing school, since younger nurses (recent nursing school graduates, one assumes) were no more comfortable with computers than were their older colleagues.

Improving nurses' confidence and their abilities to use medical devices may also be a matter of allowing them more time to learn to use the devices; 66 percent of the respondents indicated that their workload gets in the way of mastering the use of medical devices. A possible solution to this problem may be for nursing managers to establish periods of time for nurses to study the operation of the medical devices in their hospital's inventory. Manufacturers might support this effort with refresher in-service presentations, augmented by help lines (like those operated by software companies) and specially developed training materials, such as videotapes or even multimedia software.

Balancing Usability and Complexity

Manufacturers are often driven by market pressures to add extra features to a medical device, thereby disregarding the user-interface designer's rule of thumb: 20 percent of a device's functions will be used 80 percent of the time. The resultant proliferation of features is particularly pronounced in the cases of advanced electronic and computer-based medical devices, because adding extra features is relatively inexpensive, requiring only a few more lines of software code rather than additional hardware components.

How do nurses feel about the extra features? Survey responses suggest that manufacturers have already reached a limit in terms of the number of features with which nurses can cope: 33 percent said that advanced electronic and computer-based medical devices have too many features, 56 percent said they have the appropriate number of features, and 11 percent said they do not have enough features.

When asked what advice they had for manufacturers making the trade-off between keeping a product simple and providing optional features, 41 percent of the nurses said to include only a few optional features, 41 percent said to include most of the optional features, and 18 percent said to include all possible optional features. The message from nurses to manufacturers seems to be:

> Give us the features we will use and leave out the rest.

Of course, this preference conflicts sharply with manufacturers' inclination to pack products full of features; a means to increase competitiveness in feature-to-feature comparisons among competitive products.

Consistency of Medical Device Operations

The survey determined that nurses place a high value on consistency in the operation of medical devices. Regarding the more advanced devices, only 11 percent of the respondents feel that control and information-acquisition methods are "very consistent." Most feel there is room for improvement: 64 percent feel that device control and information acquisition is "somewhat consistent," 11 percent feel it is "neither consistent nor inconsistent," 11 percent feel it is "somewhat inconsistent," and only 3 percent feel it is "very inconsistent."

The solution to inconsistencies, a rich source of human error, may be the development of additional national and international standards for medical device operation. Ongoing work by ASTM to harmonize the visible and audible alarm signals associated with anesthesia devices, for example, is one step in the right direction.

Priorities in User-interface Design

As in other design disciplines, user-interface design requires many trade-offs. For example, a device that is initially easy to learn to use may not be the easiest to use in the long run. As shown in Table 4.1, the nurses' average ratings of 10 usability attributes suggest that ease of learning is the key attribute. (Learning may be defined as acquiring an understanding of what a product can do for the user and what steps the user must take to use it.)

The underlying notion may be that learning to use a device is essential to getting the job done (recall that nurses project a "can do" attitude), whereas physical comfort is not. The respondents' preference for initial ease of use over long-term ease of use provides additional evidence that nurses want

Table 4.1. *Average ratings of the relative importance of usability features of medical devices, based on a 10-point scale.*

Attribute	Importance
Ease of learning	8.0
Initial ease of use	7.4
Speed of tasks	7.2
Long-term ease of use	6.8
Minimum opportunity for error	6.1
Sense of control over device functions	5.2
Tolerance for operator erros	5.2
Minimum reliance on learning tools	4.3
Comfort of physical interactions	3.6
Aesthetic appearance	1.2

their initial interactions with new medical devices to be smooth ones.

Conclusion

Although the results of this survey were compiled from a small sample of nurses not selected at random, several themes emerge. First, critical-care nurses are very much concerned about the usability of medical devices and feel that manufacturers should place additional emphasis there. A significant number of the nurses surveyed would like a role in the development of future devices. Second, the key variable in the matter of usability is the learning process. Nurses' workloads do not always permit them enough time to master the use of a product. This fact suggests that manufacturers should pay close attention to how nurses and others learn to use products so that manufacturers can help users develop the most important skills as quickly as possible. Third, manufacturers and standards organizations need to find ways to instill greater consistency in the user interfaces of medical devices, particularly as more devices incorporate a computer-based interface. Fourth, nursing education curricula should incorporate computer training to better prepare nurses for the high-tech world in which they will work. Manufacturers can support this effort through in-service training and additional learning aids.

Chapter

5

Preventing Human Error Through Improved Design

Major league shortstops, concert pianists, and airline pilots do it, not to mention the best surgeons. Once in a while, even though they may be experienced and vigilant, they commit errors. Human beings simply do not perform tasks with the precision and repeatability of robots. As a result, workplace activities are peppered with errors, most of which are minor and can be corrected before any harm comes to pass. Unfortunately, more serious errors occur that cannot always be reversed. An article in *Anesthesiology* reports,

> In a 1985 survey of nearly 300 private practitioners across the western United States, 24% admitted to committing an error in anesthetic practice that had lethal consequences (Weinger et al. 1990, 995).

Frequently, investigations into events involving serious errors determine that human factors deficiencies associated with a particular device led to the error and that better designs could have eliminated the opportunity for error or helped with detection and facilitated a rapid recovery. Recent studies conducted by the FDA and The Ohio State University's Department of Anesthesiology reinforce this finding. The message to manufacturers is clear: Check for usability problems and opportunities for human error during the design process, not after users experience problems.

Slips and Mistakes

Two fundamental types of human error are slips and mistakes. A slip is a case of automated behavior gone astray and can assume many forms, some of which can be quite amusing. Usually a slip is quickly detected because associated behaviors are recognized as inappropriate. For instance, after taking the patient's temperature, a nurse removed the disposable cover from the sensor and threw the digital thermometer in the trash. A mistake is the product of conscious but incorrect decision making. Compared to slips, the deliberateness of mistakes makes them harder to detect, since the associated behaviors seem appropriate at the time. For instance, using a new patient monitor for only the second time in an actual case, the anesthetist pressed the button that turns off all alarms when his intention was simply to silence them for 3 minutes.

Research on human error suggests that many factors can increase the chance of error, such as inadequate work space and work layout, poor environmental conditions, inadequate human engineering design, inadequate training and job aids procedures, poor supervision, and stress. For instance, one study of human error suggests that the chances of an error increase tenfold when novices perform routine tasks in the presence of high stress and fivefold for skilled (practiced) individuals. In another study of individuals performing highly proceduralized tasks, researchers estimated that

> the probability of omitting one task out of five decreases by a factor of more than 100 when written procedures are used, compared with responding to oral directions alone. Procedures with checkoff provisions reduce error potential by an additional factor of three (Salvendy 1987, 224–225).

Magnitude of the Problem

Table 5.1 lists 20 of the most error-prone products within the FDA's purview. The data (compiled in 1991) come from the MDR system, which is administered by CDRH. The backbone of the system is 21 CFR 803.24, enacted in 1984, which stipulates:

> Manufacturers and importers must file an MDR report when their device . . . may have caused or contributed to a death or

Table 5.1. *Medical devices with as high incidence of human error.*

Device	Rank	No. of Reports
Glucose meter	1	675
Balloon catheter	2	412
Orthodontic bracket aligner	3	309
Administration kit for peritoneal dialysis	4	291
Permanent pacemaker electrode	5	278
Implantable spinal cprd stimulator	6	263
Intravascular catheter	7	262
Infusion pump	8	250
Urological catheter	9	248
Electrosurgical cutting and coagulation device	10	237
Nonpowered suction apparatus	11	226
Mechanical/hydraulic impotence device	12	213
Implantable pacemaker	13	197
Peritoneal dialysate delivery system	14	192
Catheter introducer	15	187
Catheter guidewire	16	177
Transluminal coronary angioplasty catheter	17	176
External, low-energy defibrillator	18	166
Continuous ventilator (respirator)	19	159
Contact lens cleaning and disinfecting solutions	20	158

Source: FDA, Office of Training and Assistance, Division of Professional Practices, Medical Branch, 1991.

serious injury, or has malfunctioned and, if the malfunction recurs, is likely to cause or contribute to a death or serious injury (FDA 1990, 1).

By late 1991, after more than seven years of MDR system operation, the FDA had received more than 133,000 reports. The reports are a tremendous resource for determining trends in medical device hazards, although the data are not considered statistically reliable for scientific studies.

FDA staffers review MDR reports to determine the cause of reported problems, the risk to the public, and any appropriate action. The FDA closes a report only when a problem has been resolved to the staff's satisfaction. According to Jay Rachlin, chief of the Office of Training and Assistance, Medical Branch,

> If you look just at the [more than 45,000] closed reports, 21% [more than 9000] have been closed with a finding of user error. However, it appears that as many as 60% of all problems reported involve human error and many problems attributed to mechanical malfunction involve human error. But the evidence of human error is hard to come by.

A conservative interpretation of the data suggests that errors in using medical devices cause an average of at least three deaths or serious injuries, and probably more, everyday.

MDR reports tend to be terse summaries of a particular problem, excluding the cause. Here are two examples (slightly sanitized to assure individual privacy) from the MDR reporting system for August 14, 1991.

Glucose Meter

Nursing home called [manufacturer] to ask for service on meter. Upon questioning [of caller], manufacturer discovered patient had reading of 390 mg/dl. Test was repeated and same reading achieved. Patient [was] taken to hospital, and lab got reading of 1300 mg/dl. Staff developer of nursing home ran control [a fluid sample with known glucose level] after checking meter's standard strip, which was in range. Meter read 39 mg/dl. Manufacturer asked how was strip inserted. Reply was "Pad was facing me," which is the incorrect way. Staff developer reran strip and received reading of 159 mg/dl. She admitted she taught her staff the wrong way.

Contact-Lens-Care Product

On [date], patient used device and did not use the catalytic disc to neutralize the solution. Upon insertion of lens into right eye, patient experienced pain and discomfort. Eye became swollen. On [date], patient visited doctor. Doctor diagnosed chemical conjunctivitis and prescribed vasocidine ophthalmic drops for use three times a day for five days. Doctor scheduled patient for follow-up visit, but patient did not return. Doctor indicated patient's eye would heal.

According to Rachlin both problems are likely due to human error, although these reports remain open at the present time. Hundreds of similar reports related to glucose meters and lens-care products prompted the FDA to sponsor separate human factors studies of how people interact with these devices.

FDA Conducts Studies of Human Error

Study of Glucose Meters

The FDA's glucose meter study began in 1988; Dr. Richard Kelly of Pacific Science and Engineering Group (San Diego, CA) served as principal investigator. Initially, Kelly and his associates reviewed error reports and conducted controlled observations of people using glucose meters. He reports:

> When novice test subjects used a device for the first time, we observed lots of errors of substitution, such as pressing the wrong button. However, these types of errors tended to go away with practice. The more persistent errors involved lengthy procedures and steps that required qualitative judgments, such as determining how much blood to apply to a test strip and deciding how long to leave it on the strip.

Kelly determined that the incidence of human error associated with glucose meters would likely be reduced by automating certain device functions in order to avoid lengthy procedures and to avoid steps requiring a qualitative judgment.

Subsequently, several manufacturers redesigned their devices so that users do not have to perform the error-prone task of wiping blood from the test strip. Also, the timing between sequential steps is now less critical. These are important advances, considering that the user may be feeling ill (i.e., hyper- or hypoglycemic) and have a reduced capacity to perform a lengthy task demanding high concentration.

In cases where qualitative judgments must still be made by the user, Kelly suggests the following:

> Manufacturers should make people aware of the importance of their qualitative judgments. People have to understand that they are performing a task normally performed by trained professionals and that it requires great care.

Kelly suggests educating users by means of special training and well-designed instructions.

By late 1991 glucose meters still ranked number one on the list of MDR reports related to human error, with 675 closed reports. Because the rankings are cumulative since 1984, this high ranking will persist for some time to come. However, there has been a marked improvement in the human factors suitability of these devices, including simpler display and controls and automation of the difficult steps. In a few years enough new devices will be in use to judge the contributions of improved user interfaces to the successful use of glucose meters. The apparent success of the glucose meter study led the FDA to launch a similar study of contact-lens-care products.

Study of Contact-Lens-Care Products

Contact-lens-care products also made the FDA's top-20 list of error-prone medical devices, a sure concern for the tens of millions of contact lens wearers. Problems with contact lenses occur because the porous lens material attracts microorganisms and other matter that can lead to eye infections. The FDA's initial analysis of MDR reports suggested that the instruction booklets accompanying the lens-care kits were inadequate. Accordingly, the FDA conceived a project to study the relationship between the quality of instruction booklets and the quality of lens care. Once again, Dr. Kelly contributed to the study.

Kelly and his associates reviewed a large sample of instruction booklets and found them to be

> generally acceptable in terms of reading grade level and typography—the basics. However, we found wide variation in terms of the extent and format of procedural information and warnings.

The researchers were particularly alarmed when they observed people using existing booklets to perform lens-care activities. Kelly reports:

> [We observed errors] for all aspects of lens care, including failing to understand the correct procedures, failing to follow instructions, improperly maintaining lens-care system components, and mismatching components from incompatible care systems (Kelly 1991, 103).

A test involving 48 individuals with experience caring for soft contacts showed that 37.2 percent made errors caring for their lenses. When researchers asked the same individuals to read a generic instruction booklet and repeat the cleaning and disinfecting task, the error rate dropped a statistically significant but disappointing 3.7 percent to 33.5 percent. Researchers then investigated whether variation in the format of warnings, precautions, and procedural steps affected errors. They produced two prototype booklets: an abbreviated booklet, a concise document that provided warnings and precautions in a section separate from the procedural steps on lens care; and an integrated booklet, which integrated precautions and warnings with the associated procedural steps on lens care and included a rationale for the more specific instructions.

Experiments involving the prototype booklets, an existing manufacturer booklet, and subjects inexperienced with lens-care procedures produced promising results. Persons using the integrated booklet made two-thirds the number of errors compared to the other options, and 67 percent preferred it over the others. This result suggests that manufacturers should include a rationale for warnings and precautions (integrated with procedural steps) in their documentation as one means to increase compliance and protect against errors.

Analyses of the errors committed by experienced soft contact lens wearers identified four particularly error-prone steps in the lens-care process:

1. Neglecting to wash hands prior to lens cleaning

2. Rubbing the lens an insufficient length of time

3. Failing to turn the lens over to rub its second side

4. Rinsing the lens an insufficient length of time

Kelly's study surmised:

> Each step consumes more time than seems appropriate to
> many people. A tendency to shorten or eliminate [steps] is,
> therefore, understandable.

This finding suggests that manufacturers should carefully evaluate the time requirements of particular tasks to determine
whether they stretch people's patience and can be shortened.

Tracing Critical Incidents

David Woods, Ph.D., is an associate professor in The Ohio State
University's Department of Industrial and Systems Engineering
(Columbus, OH). Over the past few years, Woods and his coau-
thors at The Ohio State University College of Medicine, with
sponsorship provided by the Anesthesia Patient Safety
Foundation, have studied the effect that human factors deficiencies in medical devices have on safe medical practice. Woods
says:

> We sought a link between deficiencies in design, the behavior
> pattern of anesthesia providers in the operating room, and criti-
> cal incidents occurring during a case.

One of their studies targeted the user interface of a specific
brand of infusion controller, a microprocessor-based device that
delivers anesthetic drugs to patients undergoing surgery.

> We looked at an infusion controller that exhibited many of the
> flaws [human factors expert Donald] Norman says will lead to
> problems. It has a complex and arbitrary sequence of opera-
> tions, hidden functions [functions embedded under layers of
> software], ambiguous alarms, poor feedback in response to
> control inputs, and poor mappings between user actions and
> the state of the device. Not only do these flaws lead to errors,
> but they make it hard for users to detect and correct an error.

When Woods and his associates observed the device used
during cardiac surgical cases, they noted how it negatively
affected clinicians' work. They found that many users had difficulty performing setup tasks, which included more than 20 separate steps. Furthermore, because the device may be set up
several hours before a surgical case, setup errors might not be
detected until they create inconvenience and perhaps danger
to the patient. Regarding the use of the device during a surgery,
Woods reports:

Users confused indications of demanded [fluid] drop rate with measurements of actual rate, and the lack of feedback and the presence of hidden modes combined to hide device state and behaviors from the physician (Cook et al., pending).

With the cooperation of anesthesiologists practicing at The Ohio State University Hospital, Woods and his associates were able to document so-called critical incidents involving the infusion controller within hours of the event. A sample critical incident follows:

During the cardiopulmonary bypass pump run, the practitioner attempted to start the [sodium nitroprusside] (SNP) drip to control hypertension. Shortly after seeing a drip rate and starting the SNP drip controller, a "no flow" alarm began. The practitioner briefly examined the intravenous tubing sets above the controllers and found a manual thumbwheel device in the closed position. The thumbwheel was opened and the SNP controller restarted. Shortly after restarting the drip the no flow alarm recurred. The practitioner examined the bags of fluid and found the dobutamine drip running wide open, although the dobutamine controller was powered down. Trace of the intravenous tubing path showed that the SNP tubing had two occlusion clips placed in series. The first was placed in the controller designated for SNP and the second in the controller for dobutamine. The intravenous tubing for the dobutamine drip did not have any occlusion clip connected.

Woods' team documented three other critical events during a three-month period. Each incident involved the inadvertent delivery of a drug when the delivery device was supposed to be turned off. Woods attributes the critical incidents to the complexities of the infusion controller, which deprives practitioners of a clear picture of the state of the infusion process, leaving practitioners to develop coping strategies. Rather than calling the value of automated infusion controllers into question or indicting one manufacturer's device, Woods hopes that his team's research leads manufacturers to develop more-usable products.

Taking an Active Approach

Implementing usability testing at several points during the design process is a productive way to detect errors and identify opportunities for design improvement (see chapter 5). A test involving 10 or more individuals using the same product normally reveals a plethora of recurrent errors. Since people rarely make the same random error, the recurrent errors indicate a flawed design. The flaw may be specific, such as a confusing label, poor control positioning, or a missing display message. On a more universal level the flaw may simply involve expecting too much of people (e.g., placing too high a demand on their memories, requiring that procedural steps be performed with precise timing according to a pace set by the device, or providing inadequate feedback to determine if a task has been performed correctly or not). Regardless, both specific and universal flaws can be exposed in usability testing.

What can be done to develop designs that will fare well in a usability test? The answer rests in dozens of technical papers and textbooks on human factors engineering. As a start, Donald Norman offers four guiding principles:

1. Make it easy to determine what actions are possible at any moment (make use of constraints).

2. Make things visible, including the conceptual model of the system, the alternative actions, and the results of the actions.

3. Make it easy to evaluate the current state of the system.

4. Follow natural mappings between intentions and the required actions; between actions and the resulting effect; and between the information that is visible and the interpretation of the system state (Norman 1988, 188).

References

Cook, R., D. Woods, and M. Howie. Pending. Unintentional delivery of vasoactive drugs with an electromechanical infusion device. Manuscript submitted for publication in *Journal of Cardiothoracic Anesthesia*. Contact David Woods for specifics about the incident cited.

Kelly, R. 1991. Assessing user compliance with procedures for soft contact lens care. In *Proceedings of the Human Factors Society 35th Annual Meeting*. Santa Monica, CA: Human Factors Society.

Medical device reporting questions and answers. 1990. Rockville, MD: FDA/CDRH.

Norman, D. 1988. *The psychology of everyday things.* New York: Basic Books.

Salvendy, G., ed. 1987. *Handbook of human factors.* New York: John Wiley & Sons.

Weinger, M., and C. Englund. 1990. Ergonomic and human factors affecting anesthetic vigilance and monitoring performance in the operating room environment. *Anesthesiology* November:995.

Determining User Requirements

Chapter

6

Conducting Effective Focus Groups

In medical device development the current trend is toward "user-centered design," an approach that emphasizes early and continual user involvement. One method to achieve this is through the use of focus groups, interviews with prospective users conducted in a group setting. Focus groups are an effective tool to define product requirements, clarify user preferences, obtain feedback on alternative design concepts, and validate a final design concept.

Because of the dynamics of group interaction, the information obtained from focus groups is more useful and comprehensive than that obtained from a series of interviews. The goals of a typical focus group are to freely explore design issues at hand, produce fresh insights, and reach consensus on major design trade-offs.

Benefits of Focus Groups

Manufacturers can realize a number of important benefits from learning how to conduct effective focus groups and making them part of an overall usability engineering effort. The immediate benefit of focus group research is that it puts designers in touch with users. By listening to users articulating product needs, preferences, and concerns, designers can strip away their own biases and produce more empathic and useful designs.

As one might hope, focus groups often confirm usability specialists' expectations regarding desirable design features. However, almost every focus group leads to one or more important unexpected findings. For example, participants might suggest eliminating a product feature, such as a clock on a menu display, that designers had thought users valued. Or, a focus group member might suggest a new feature, such as a connection for an IV pole, that could make a product more portable and differentiate it from competing products. In these ways focus group participants are a rich source of ideas for product design and improvement.

Another benefit of conducting focus groups is that they enable researchers to collect feedback from a large number of users in a short time. Researchers meeting with 2 focus groups per day, for example, can communicate with as many as 100 users relatively quickly.

Conducting the Focus Group

The time spent actually running focus group sessions (a total of perhaps six to eight hours over three sessions, for example) represents only a fraction of the time invested. To be most productive, focus groups take considerable planning and follow-up reporting. The basic steps in conducting focus groups are defining information needs, writing a script, preparing exercises, recruiting subjects, preparing facilities, running the sessions, and documenting the results.

Defining Information Needs

Effective focus groups call for a commitment to objective, professional research. As one might expect, the quality of a focus group hinges on the quality of the questions asked. High-quality questions arise from a need for specific, practical information, such as information that will help you decide to use either mechanical or snap-dome, membrane switches in a device interface. Focus group participants may confirm the need for control panels that are impervious to fluid spills and easy to clean, suggesting membrane switches, forsaking the superior tactile feedback afforded by so-called mechanical keys. Therefore, you should ask only those questions that provide useful information. Furthermore, it is a good idea to meet with design team members to solicit their information needs before you write the questions. Most of your questions should focus on product usability; however, pragmatism may dictate that you ask a number of questions that reflect marketing or engineering concerns, so that you or they do not have to bear the cost of a separate round of focus groups.

Writing the Script

Group discussions progress more smoothly if you prepare a script to guide them. It might be 5 to 10 pages in length, listing your questions in their anticipated order. Organizing the questions by topic helps to structure the discussions. Questions should be written in an informal, conversational style. Once you have a draft script (ideally written by one person for the sake of continuity), distribute it to the design team members for comment. You might also want to schedule a meeting to discuss changes. Be sure to guard against scripts that stress one designer's concerns over those of others.

One approach to script organization is to discuss the full range of general issues first, then move on to specific questions. This method helps participants build a solid understanding of basic design issues and develop a sense of their own design priorities. Some researchers also believe this approach helps people follow the conversation more easily and participate more. An alternative approach deals with one issue at a time, moving from general to specific discussions. This approach is less redundant, but makes it harder for participants to make design trade-offs, since they lack a sense of the overall issues.

Proofread all script questions to make sure they make sense and are worded in an unbiased fashion. Otherwise, the results of the study may be misleading and the study itself may be discredited. Questions should have a neutral tone, but should involve the participant. Examples might be questions such as the following:

- How do you feel about carrying around a heavy product?

- How does the appearance of the product compare to your expectations?

- Can you identify any advantages or disadvantages of using a touch screen?

- Can you suggest opportunities to improve the cable harness?

A focus group should last 2 to $2^1/_2$ hours, with a 5- to 10-minute break in the middle. After the second hour participants' energy levels begin to wane, so the script should not include more questions than can be covered in the time available. Protracted discussions can be avoided by setting time limits for each script section, building slack into the schedule, and designating certain questions as optional, to be asked only if time permits.

Most researchers feel that conducting one focus group per day can be handled with little difficulty, but that holding two is strenuous. An international marketing manager for Siemens Medical Systems conducts at least two focus groups for each stage of product development that requires evaluation, but she limits focus group sessions to one per day.

> The time between groups lets us digest the information we receive and adjust our approach if we aren't getting the information we need.

For companies with little or no experience in conducting focus groups, she suggests running a test group to make sure interviewers are able to obtain the desired information, before conducting a large series of focus groups or spending a lot of money on testing conducted away from the company's principal facility.

Preparing Exercises

A focus group does not have to be restricted to discussing issues. In fact, a productive group might include several exercises. For example, group leaders might administer a questionnaire about design priorities, conduct a hands-on exercise with a prototype device, demonstrate a product's special features, show a videotape of the product in its intended-use environment, or have participants complete a comparative rating sheet after reviewing three design options. Such exercises can enhance subsequent discussions and also provide a break in discussions when people are at the limit of their attention span.

Pragmatic researchers should not hold themselves up against academic preconceptions of how a focus group should progress. More important than sounding smooth and professional, the focus group simply needs to provide researchers with the information they seek to move forward with the design.

Recruiting Participants

Recruiting focus group participants is a double challenge. First, you have to decide what type of people to recruit, then you have to find them and convince them to join the group. The international marketing manager at Siemens quoted above began by using a survey to screen potential focus group participants. Then, based on the results, she invited 30 people to participate in focus groups on a continuing basis, convening the groups whenever the need for information arose.

Medical device users tend to be quite responsive to focus group invitations, perhaps because they are enthusiastic about their work and have a strong interest in seeing medical devices improve. Therefore, it is important to explain the basic nature of the research to prospective participants as a means of piquing their interest. Researchers can offer an honorarium equal in value to the time participants will spend in travel and attending the focus group, but it should not be presented as the major benefit. Researchers who have a distaste for recruiting, or do not have the time to do it, may resort to using recruiting firms to obtain focus group members. A last-minute solution for recruiting people in specific occupations, such as nursing, is to approach temporary worker agencies that offer a wide range of qualified personnel at a reasonable rate.

Focus group participants should represent a good cross section of users. Researchers can control the mix of participants so that commonalities and differences emerge in the views of individuals representing homogeneous or heterogeneous user groups. If the product you are evaluating is an electrocardiograph, for example, the intended users might include emergency medical technicians, registered nurses, physician assistants, internists, and cardiologists, perhaps two from each occupation. Or, you might recruit eight participants from each occupation and conduct five homogeneous sessions.

How do you decide between conducting heterogeneous and homogeneous groups? Experience suggests that heterogeneous groups are the appropriate choice when you want a variety of viewpoints to stimulate discussion about basic design concepts you wish to explore. By way of contrast, homogeneous groups are usually best for evaluating detailed designs, since participants with a common background will be better able to establish a rapport and work toward a meaningful consensus (Calder 1971, 353–364).

Occupation is only one of many possible selection criteria. Others include age, sex, experience working with high technology, working environment (clinic, community hospital, or teaching hospital, for example), and familiarity with specific products. Again, remember that the goal is to achieve a good cross section of users. Marketing personnel may be able to help define a user profile that can form the basis of recruiting requirements, and provide good sources of leads for possible focus group participants. However, be sure not to stack the focus group with loyal customers who may feel indebted to their host and, therefore, hesitate to be critical.

Focus groups usually consist of 6 to 12 participants. The appropriate size depends on the scope of your project and moderator preference. The international marketing manager at Siemens initially worked with larger groups, but now prefers groups of 4 to 6 participants.

> Smaller groups are more energetic. People show more creativity and play their ideas off one another.

To avoid misleading results, she feels it is important for companies such as Siemens, which develops products for an international market, to conduct focus groups in different regions of the country (and world) rather than just locally. She has observed, for example, that

> potential users on the West Coast are more open to innovation, while people on the East Coast and in Europe are more conservative.

The recruitment effort is not complete even after all the participants have been enlisted, because no-shows still pose a risk to the success of focus groups, especially if the group is composed of people from varying occupations. One way to protect yourself against no-shows is to ask people to provide several days notice of cancellation, to send clear travel directions in a confirmation letter at least one week prior to the focus group, and to place a reminder call a day or two before the event.

Preparing Facilities

Focus groups can be conducted wherever there is a quiet meeting space that can accommodate up to 12 participants, the researchers (normally 1 to 3 people), and any required equipment. In many areas of the country, there are companies that specialize in renting out facilities equipped with a one-way mirror and videotaping equipment. Such companies may also handle recruiting and provide refreshments for participants. Another solution is to rent a small function room at a hotel. This may be the best approach if you find that conducting a focus group in the participants' work environment (e.g., the nurses' conference room) or at the developer's headquarters would make people feel inhibited or jeopardize confidentiality.

Participants seem most comfortable when sitting around a conference table, although a circle of chairs is acceptable. A table may be essential if participants will be asked to complete a lot of paperwork. Everyone participating in the session should wear a name tag so people can converse on a first-name basis. Refreshments can be placed in a corner of the room.

Running the Sessions

Two hours may sound like a long time to talk about design issues, but the time tends to fly by. One way to make the most of the time available is to ask participants to arrive 10 to 15 minutes early to take care of administrative details. Offering refreshments or even a meal before the session starts can motivate people to arrive on time.

It is best to have one person lead the focus group—someone who has experience as a moderator, is able to maintain objectivity, has a natural ability to establish a rapport with participants, and has public speaking skills (Goldman 1987, 145–154). The lead moderator's job is to introduce topics for discussion, have all participants to respond with candor, encourage debate, help the group reach a consensus when appropriate, and keep the meeting on schedule. When covering highly technical issues, it is helpful to designate an assistant moderator who can ask follow-up questions and provide technical clarification.

The moderator should behave in a calm fashion as a way of encouraging the participants to do the same. Speaking casually and using humor sets the right tone for candid and energetic responses from the participants. The introduction is the best time to set this tone. The moderator should welcome the participants, thank them for taking the time to attend, explain the mechanics of the focus group, and provide some ground rules, such as the following:

- Everyone will have a chance to speak.

- People should not interrupt each other.

- People should not criticize the views of others.

Next, the moderator can introduce the major topic of discussion and launch into the scripted questions. A talented moderator will paraphrase questions rather than read verbatim. He or she will avoid biased questions (e.g., Do you think the model has a modern-looking shape that will make it easy to hold? Instead, the question should be worded: How do you feel about the shape of the product?). A well-informed, skillful moderator will know how to direct participants in ways that produce useful, unanticipated results, as illustrated by the following dialogue from a hypothetical focus group:

Moderator: At this point, we have discussed the various types of information that might be contained on a computer-based anesthesia record. Now, let's discuss how you might enter information into the record. For example, how would you enter a drug administration?

Participant: Assuming there is a form presented on the computer screen, I would enter the drug name, how much I gave the patient, and when I gave it.

Moderator: Can you be more specific about the physical actions you would perform?

Participant: Well, I've seen some systems that use a touch screen, although I don't know much about them. I think you just touch parts of the screen to select or specify information to add to the record.

Moderator: How do you feel about interacting with a touch screen?

Participant: I've never liked the feel of a touch screen, although I guess I could get used to it. Plus, you need a lot of space for the touch areas, which often obscures information on the record. I would like to be able to write directly on a form—the way I do today—and somehow have the information get into the computer.

Designers or other product development personnel may choose not to participate directly in focus group discussions. They may view the proceedings through a one-way mirror or via live video, or they may simply watch videotapes later. The advantage of live participation, however, is that it provides the opportunity to obtain clarification of participants' comments.

The moderator should encourage people to respond spontaneously, and should direct questions toward quieter individuals if necessary. Often, one or more participants begin to dominate the discussions. In such cases the moderator needs to use tact while redirecting the discussion to others. Otherwise, the focus group findings will be skewed toward the views of only a few.

When seeking a consensus, it is helpful for the moderator to summarize the participants' final position on an issue, then ask the group if they concur with the summary. For example, the moderator might say,

> I think you are telling me that you are willing to carry around a six-pound patient-data terminal, but that the battery needs to last for at least a normal, eight-hour shift. Is that accurate?

Near the end of the session, the moderator should give participants a final chance to comment on issues discussed earlier or ones that might not have been raised. Then the moderator should thank the group members for their participation and point out the importance of their feedback to the design team. If confidentiality is important, participants should complete a confidentiality agreement prior to starting the focus group and be reminded of it as they depart.

As participants depart, they should be compensated for their time. Normally, the host company will require participants to sign receipts for the compensation. If you want the participants to return at a future date, this is a good time to have them complete an index card indicating whether they are interested in further involvement.

Documenting the Session

Videotaping the focus group enables people who were unable to attend the session to review the proceedings. It also enables researchers to review their work, and thereby to improve their techniques (for example, by watching for and eliminating bias in the questions). One video camera is usually sufficient. Some companies with dedicated facilities for focus groups mount the camera behind a one-way mirror so that it is inconspicuous. Others place it high in the corner of a room to capture all participants. The production quality of the tape will be enhanced by employing a camera operator to capture close-ups of people talking. Dedicated facilities typically incorporate a ceiling microphone; an alternative is to place one or more microphones on a table at which participants are seated.

Usually, a research assistant serves as the primary note taker, recording important portions of the discussions. The moderator might also take notes. Sometimes, companies produce a complete transcript of the proceedings, using secretarial support or firms specializing in the service.

Reporting the Results

Normally, focus group findings are documented in a 10- to 20-page report, which may include an executive summary of findings, an overview of the focus group method and participants, conclusions, and appropriate attachments (compilations of questionnaire and exercise data, for example). It is convenient to base the focus group report on the way the script is organized—the relevant findings (answers to questions, comments, survey data) can be presented following each scripted question.

If you conduct a series of focus groups, it is unlikely that management will be willing to watch all the videotapes. You may want management to obtain a firsthand feel for the proceedings, however. One solution is to create a 5- to 15-minute highlight tape capturing important or controversial participant comments. The tape might also include a series of responses by different groups to the same question, thereby revealing patterns or trends.

The Importance of Timing

The value of a focus group can be enhanced by good timing. If conducted during the concept phase of a product development project, an effective focus group can help to differentiate promising design concepts. But the information may have limited value or may even be unwelcome if it comes late in the design process, after key design decisions have been made. Design managers rarely react well to focus group findings that contradict irreversible design decisions.

In other words, product developers should reserve the use of groups for times in the design process when the design team wants to learn more about users, needs help conceptualizing a product, or wants to develop a consensus on the suitability of one or more designs. Ken Gary, manager of systems engineering at Ciba Corning Diagnostics Corp. (Medfield, MA), which manufactures blood chemistry analyzers, endorses the focus group technique as a means to guide user-interface development.

> Focus groups are an appropriate technique to apply in user interface development when you are evaluating new product ideas and when you want to get user reactions to a detailed design, before detailed engineering.

However, many development budgets preclude an extended series of focus groups, considering that the cost of a focus group can be $2000 to $4000 for the first session and $1000 to $2000 for each additional session, not including travel costs. Many companies may budget for just a single series of three focus groups conducted locally. In this case the best time to conduct focus groups is probably during design conceptualization, before any serious design takes place, or at the point when the development team could introduce three basic concepts to participants for their critique. Later in the design process, a focus group might be helpful as a means of validating a detailed design, presented in the form of a working model or electronic prototype. However, considering that the basic design may well be established at that point, a usability test with perhaps eight subjects might be a more productive approach to obtaining user-interface design feedback (see chapter 23).

Conclusion

Conducting focus groups is not a simple matter of getting people in a room and chatting about design issues. Such casualness introduces the risk of obtaining only partial or misleading feedback from users because of bias in the questions posed or slipshod research techniques. Instead, a productive session involves plenty of advanced planning, not only to clarify key design issues, but also to determine the best point in the design process to conduct focus groups.

The concept development stage is likely to be the best time to conduct focus groups if one's usability research budget is limited. Ideally, focus groups can be repeated as a design progresses from loosely understood requirements to a preproduction design. Findings from iterative sessions, combined with the results of other usability research efforts, enable midcourse design corrections that move a device design in the right direction.

References

Calder, B. 1977. Focus groups and the nature of qualitative marketing research. *J Marketing Res* 14:353–364.

Goldman, A., and S. S. McDonald. 1987. *The group depth interview: Principles and practice.* Englewood Cliffs, NJ: Prentice-Hall.

Chapter

7

The Making of a Task-Oriented Product

Sometimes product developers focus too much on technological robustness and too little on usability. This is how the world has become full of products that look impressive but are hard to use. Yet, technological robustness and usability need not be mutually exclusive product attributes. Up-front analysis of how people will use a product can lead to user interfaces that are both task oriented and feature filled. Here, the concept of a task-oriented user interface is introduced by means of a comparison to an unlikely product—the auto mechanic's tool chest.

Almost everyone who has visited an automobile repair shop has seen the red enamel tool chests—one per mechanic—that stand close to the engine compartments of cars being serviced. To someone concerned with how people do their jobs, these tool chests are quite revealing.

A good tool chest can be quite expensive and usually holds a selection of tools many times more valuable than their container. Good mechanics take great care organizing their tools in the chest so they can work efficiently. Mechanics know exactly where each tool can be found, and they return it to its proper place after use. Closer inspection of the tool chest and its many drawers reveals a logical internal order that enables frequently used tools to be placed in convenient locations and keeps tools for a specific task in the same spot. Credit for the internal ordering goes partly to the tool chest manufacturer and partly to the mechanics who tailor their tool chests to suit their work patterns and preferences.

A tool chest does many things to facilitate car repair. First, it reduces workload by enabling a mechanic to quickly and instinctively open the correct drawer to find the appropriate tool, rather than sort through a tangled-up pile of tools. Second, it provides cues that help get the job done right. For instance, a drawer might be reserved for essential tune-up tools, reminding the mechanic to perform the tasks associated with each one. Third, the tool chest reduces workstation clutter so that the mechanic can concentrate on the task at hand. Fourth, the tool chest is flexible enough to allow mechanics to arrange their tools to satisfy their individual needs. In short, a tool chest is a task facilitator—a device that helps a mechanic repair your car.

Why is the auto mechanic's tool chest relevant to the development of medical products? It serves as a metaphor for the complex user-interface structure of a product such as an integrated anesthesia monitor. This product can be thought of as an electronic tool chest containing specific tools to monitor a patient's heart function, respiration, temperature, oxygen-saturation level, and blood pressure. In fact, these tools are commonly used in the form of individual single-purpose monitors arranged side by side on the shelves of anesthesia gas machines. The integrated monitor's top-level display is like the top shelf of a tool chest, which holds the tools most frequently used. In the case of the monitor, the items include vital-sign waveforms, numeric readings, and controls to actuate special functions and navigate through lower-level displays.

Secondary information is presented on lower-level displays as if it were being stowed in one of the tool chest's drawers. Bringing up displays and dismissing them by pressing keys is comparable to opening and closing drawers. Individual displays categorize and highlight information to help users locate it faster, just as drawer organizers help mechanics find a tool faster. And as with a tool chest, the internal ordering (user-interface structure) of the monitor ideally helps users complete tasks.

Users cannot, however, easily rearrange a monitor's contents to suit their needs and preferences. Instead, a medical product developer must create a user-interface structure that works well for most users. Understanding how most people will use the product is a good starting point.

Understanding the User-Product Relationship

Human factors specialists have developed a large set of task-analysis techniques that clarify the user-product relationship. There really is no mystery to task analysis, even though some people in the human factors profession make it seem complicated. Most established techniques are systematic descriptions of a person using a product. This information can be expressed in logs, charts, diagrams, and tables that answer the following questions:

- What are users thinking?

- What information do users acquire?

- Where are users looking?

- Where do users place their hands?

- At what point do users make decisions and take action?

- How frequently and urgently do users take action?

- How long do users take to complete the action?

- How well do users perform the action?

- How often do users make a mistake?

For example, a good way to tell what people are thinking is to have them speak as they perform a task and record their comments. The flow of visual information to users can be reflected in a flowchart; eye-tracking headsets can show where users are looking (Figure 7.1). Link diagrams can trace how people move their hands from control to control. Decisions and actions can be illustrated in a flowchart resembling those developed for computer programs or controlled processes. The frequency and urgency of user actions can be determined through observations or interviews and expressed in tables and charts that suggest which ones are the priority tasks. Performance measures can be

Figure 7.1. *Eye tracker system (Applied Science Labs Model 4100H) enables researchers to tell where the person wearing the head-mounted optics device is looking at any given moment.*

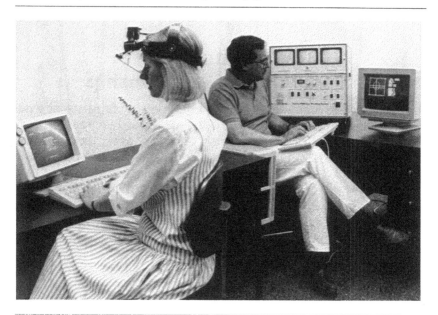

Courtesy of Applied Science Laboratories.

drawn from items such as task times, success rates, and error tallies.

A host of textbooks include tutorials on using these kinds of task-analysis techniques. Some of the better reference texts are listed at the end of this chapter. Most of the techniques can be applied whether the product already exists or is only a concept.

Consider the kind of task analysis the developer of a relatively simple medical product, such as a blood pressure monitor, might perform to clarify user-interface requirements. In the course of several days of research, the developer might generate the data shown in Table 7.1, using one of several different approaches.

User Observation

Researchers observe clinicians using monitoring devices in the actual use environment, recording the number of times that

Table 7.1. *Analysis of tasks associated with a hypothetical blood pressure monitor.*

Tasks	Task Frequency	Task Urgency
Start/stop manual cycle	High	High
Start/stop auto cycle	Medium	Low
Adjust cycle time	Low	Low
Determine pressures	Very high	Very high
Determine trend	Medium	High
Turn alarm on/off	Low	Low
Set alarm limits	Very low	Low
Silence alarm	Low	Very high
Perform self-test	Very low	Very low

users perform various tasks and rating the apparent urgency of the tasks. The researchers may be assisted in their observations by a clinician who can interpret the events as they unfold.

User Interviews

Researchers discuss monitoring tasks with a practical sample of clinicians, starting with 10 or so people and evaluating the variability of the feedback before going any further. Clinicians may provide general comments on task frequency and urgency, or they may be asked to give verbal ratings. This method of collecting feedback is fast and includes a personal touch that generates goodwill toward the manufacturer. Researchers must be aware of their own personal biases as they record data from these sessions.

User Surveys

Researchers collect data using a questionnaire. This approach enables a more reliable statistical analysis, although it sacrifices the one-to-one contact that promotes a better understanding of users' needs.

Assessed in the context of established usability goals (see chapter 8), such data could form the basis for decisions on how to structure the user interface and what displays and controls to provide. Based on the information in Table 7.1, for example, a designer might decide to use a 40 percent larger LCD display, which would allow users to read uncluttered displays of blood pressures and trends at the same time, instead of requiring them to press a key to toggle between displays. Users require frequent and fast access to both types of information. Spreading the information across two, mutually exclusive screens presented on a single display may frustrate users and might make it difficult for them to obtain crucial information.

A designer might also provide an oversized push button to silence the alarm. Despite the low frequency of use, the task of silencing the alarm is an urgent one and should not require a time-consuming visual search to find the right key.

Finally, it might prove useful to employ a menu key and two "soft keys" (also called function keys) to access display pages that enable users to turn the alarm on/off, adjust the alarm limits, and adjust the auto cycle. Because these adjustments are not performed frequently or with undue urgency, they can be imbedded in a display hierarchy.

The degree to which extensive task analysis is needed depends on the complexity of the product being developed. It may be unnecessary, for example, to develop wall-sized flow-charts for a simple product (such as a digital thermometer) that requires mostly sequential decisions and actions. One could simply talk to a sample of potential users, asking questions such as the following:

- How often do you use the device?

- What basic tasks do you perform with it?

- What aspects of the device make it simple versus complicated?

- Which tasks are the most important?

- Which tasks have to be done with urgency?

- Which tasks are extraneous?

- How could the device be improved to do things better?

- How could it be improved to do things faster?

To some extent, these are the same kinds of questions that marketing people ask their potential customers. It is important for user-interface designers to ask the questions directly, however, because the answers have more impact when they are heard firsthand. Also, when designers meet with users in their working environment, designers tend to feel greater empathy toward them.

An initial round of task analysis is sure to leave designers better prepared to create a task-oriented user-interface structure. If the product is simple, the analysis should be completed in a week or two. A more-complex product may require a month or more of analysis, but taking the time will put the user-interface design process on the right path.

Developing a User-Interface Structure

User-interface designers are advised to consider the "magical seven plus or minus two" rule, which suggests that people can reliably process only seven plus or minus two (i.e., five to nine) elements at a time (Miller 1956, 81–97). This is one reason zip codes and telephone numbers are divided into more manageable chunks instead of uninterrupted strings (Figure 7.2).

Some designers are fond of extending the rule to the design of user-interface structures, even though it has more to do with memory recall than with item recognition. The point is that

Figure 7.2. *Examples of "chunked" information and physical features.*

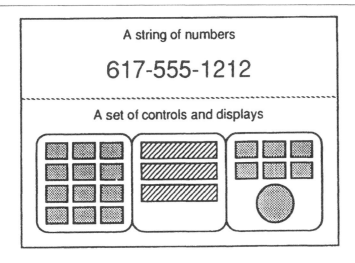

people perform tasks well when user interfaces are divided into logical elements, as long as there are not too many of them.

Following a conservative approach, the user-interface structure probably should present no more than five basic tasks to the user because more may start to overwhelm some users. Five or fewer will help users build a mental model of how the user interface is structured, giving them the sense that the product is simple to use. The ultimate goal is for users to be able to describe the basic tasks when the product is not at hand, indicating that they have established a firm mental model.

To converge on five or so basic tasks, designers should focus on the simple question: "What do people do with this product?" The answer should be drawn from the task analysis results and conversations with potential users. Initially, it may be hard to limit the list to five. A longer list of tasks, however, usually can be subdivided into related groups that comprise a more fundamental set of tasks. For a blood pressure monitor the five basic tasks might be as follows:

1.   Determine the patient's blood pressure.

2.   Set the monitor to cycle manually or automatically.

3.   Determine blood pressure trends.

4.   Identify alarm conditions.

5.   Determine if the monitor is working properly.

After the basic tasks are identified, designers can follow a similar process to determine subtasks or steps. For instance, Task 1 might be expanded into the following subtasks:

• Determine the systolic, diastolic, and mean pressures.

• Determine the "age" of the last pressure readings.

• Determine whether the pressures are increasing or decreasing.

The next step is to conceptualize a user interface that presents the basic tasks so that new users, for example, can understand quickly what the product does. This step can be relatively straightforward if the product's user interface is mostly hardware- or software-based, rather than an amalgam. It is important to note that although a hardware solution may incorporate a computer display, such a display would not present control or display options to the user.

User interfaces that are part software and part hardware offer the potential for improved usability over ones that are exclusively hardware or software based. Also, they typically enable a user interface to be more compact. Approached properly, tasks performed with high frequency and urgency should probably be assigned to dedicated hardware, while (returning to the tool chest metaphor) software provides "drawers" that contain the rest of the product's capabilities. A review of the task-analysis results, such as those presented earlier in Table 7.1, should suggest an appropriate allocation.

Designers must strike a balance between the need to expose all basic tasks and the need to keep user interaction with the product efficient. Otherwise, one or more of a product's basic capabilities might become hidden in a software "drawer." In the case of the blood pressure monitor, for example, designers might decide to provide a "self-test" key on the control panel, even though it is used infrequently and without great urgency, or perhaps a status indicator on the display to serve as a drawer label, telling the user what is inside.

The final user-interface structure could take several shapes, varying in terms of task hierarchy depth and breadth. Broader hierarchies can push the limits of the seven-plus-or-minus-two rule and make the interface seem too complex. Deeper hierarchies tend to hide basic tasks and increase the time it takes to perform a task. Designers need to develop the right blend of breadth and depth with care and intelligence. If one is forced into a deep hierarchy, for example, task speed can be enhanced by providing links (also called wormholes) between the drawers, as shown schematically in Figure 7.3. Dedicated controls can be used to switch users quickly from one drawer to another.

Before deciding on a preferred user-interface structure, designers should obtain feedback from potential users. Many developers use focus groups and user-interface prototypes to obtain this feedback because these techniques enable them to collect individual opinions as well as a group consensus.

Simplifying User Interactions

Another important way to make a product task oriented is to provide convenient mechanisms for user-product interactions (for opening and closing the drawers). For example, control panels should have hierarchical labeling schemes that lay out

Figure 7.3. *Links between user-interface elements, also known as wormholes, enable a user to perform tasks more rapidly.*

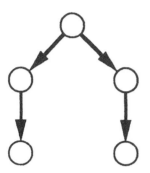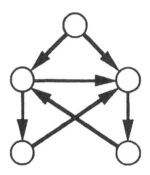

User-interface structure
without wormholes

User-interface structure
with wormholes

the basic tasks for users. Some control panels might even include a mimic diagram to direct the correct sequence of hand motions. Main menu options in software-driven interfaces should be phrased in concise, ordinary language; these terms form the titles of lower-level displays (Figure 7.4). Displays should follow the same control-action syntax both when they are brought up and dismissed and when display information is modified.

Many computerized products present a mix of control-action syntax. Sometimes, users choose an object (such as systolic pressure) and then select an action (increase upper alarm limit). At other times, users choose the action and then select the object. At a minimum, the mix causes confusion and irritation. Worse, it can lead to serious operating errors.

Although a well-labeled push button might seem out of vogue in a computer world, it is hard to beat in terms of clarity of purpose, acquisition time, and feedback to the user. Therefore, designers should not hesitate to use push buttons (or comparable hardware components) liberally, especially in conjunction with basic tasks. Usability tests indicate that people using products for the first time respond positively to such dedicated controls as long as they do not get too numerous (see chapter 23). Tests of several complex medical products, as well as consumer electronic gear and scientific products, suggest that the number of dedicated controls should be limited to 10 to 15

Figure 7.4. *An option presented on higher-level menus forms the title of an associated, lower-level menu.*

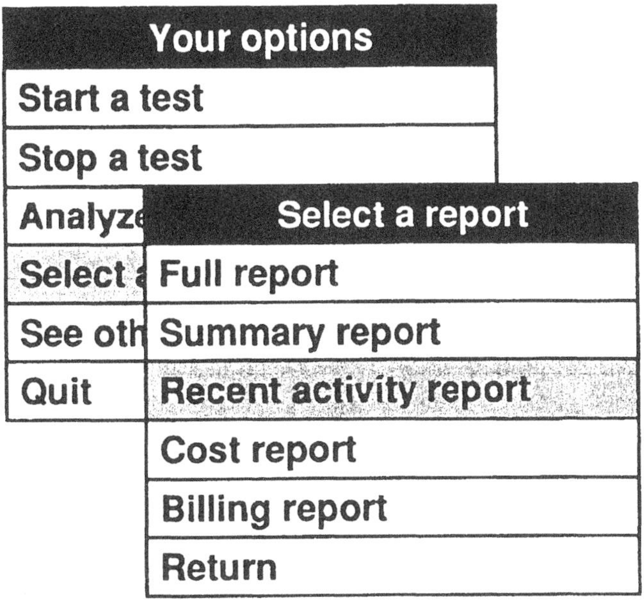

Your options
Start a test
Stop a test
Analyze
Select a
See oth
Quit

Select a report
Full report
Summary report
Recent activity report
Cost report
Billing report
Return

(divided into groups of five or so) to avoid intimidating users while at the same time exposing sufficient device capabilities. According to this rule of thumb, a numeric keypad could be counted as one control, assuming that demarcation lines or a similar highlighting technique is used to give the pad a unified appearance.

Usability tests also indicate that new and experienced users depend on learning tools such as on-line help and user manuals. Tests show that people do not like to read on-line help as much as they prefer to do things by trial and error. When users require assistance performing a task, however, they resort quickly to pressing a "help" key or flipping through the user manual. This finding suggests that developers should treat learning tools as extensions of the user interface. As a side benefit, if a product is user oriented, providing on-line help or writing a user manual is relatively easy. If you have problems structuring ways of assisting users, it may be a symptom of a more serious problem in the structure of the user interface itself.

Refining the User Interface

Once designers establish an appropriate user-interface structure and simple methods of user interaction, they can move forward with detailed design, which undoubtedly will require changes to the original user-interface structure and interactions methods. Compromises also will have to be made in response to other engineering considerations. The key, however, is to keep the vision of a task-oriented product intact. It pays to revisit the results of earlier analyses and user feedback to confirm that the vision is not slipping away.

The original vision of a task-oriented product can be undermined by poor user-interface design. Therefore, one should pay close attention to details. Usability tests show, for instance, that one poorly conceived label can lead to major usability problems. For example, the term *maintenance* on an infusion pump could be interpreted either as (1) an infusion mode or (2) an option chosen to calibrate the device. Designers also must guard against major design changes that corrupt the user interface. For example, switching from a large, graphical LCD display to a smaller, character-only display would require a complete rethinking of the user-interface structure and interaction scheme. Protection of the user interface from compromise during the detailed design phase requires perseverance by user-interface designers, bolstered by management commitment to the concept of a task-oriented product.

References

Miller, G. 1956. The magical number seven, plus or minus two: Some limits on our capacity for processing information. *Pysch Rev* 63:81–97.

Recommended Readings

Cushman, W., and D. Rosenberg. 1991. *Human factors in product design, advances in human factors/ergonomics*, Vol. 14. New York: Elsevier Science Publishing.

Kantowitz, B., and R. Sorkin. 1983. *Human factors—Understanding people-system relationships*. New York: Wiley.

McCormick, E. 1976. *Human factors in engineering and design*, 4th ed. New York: McGraw-Hill.

Salvendy, G., ed. 1983. *Handbook of human factors*. New York: Wiley.

Schultz, D. 1982. *Psychology and industry today*, 3rd ed. New York: Macmillan.

Woodson, W. 1981. *Human factors design handbook*. New York: McGraw-Hill.

Chapter

8

Setting Product Usability Goals

When technology reaches a point at which all leading products of a similar type share the same basic capabilities, smart product developers turn to other design attributes for a competitive advantage. Some developers focus on product appearance, reliability, or maintainability; others look for ways to improve their customer service program or simply do what microeconomic principles suggest—lower their price. A less common reaction is to focus on product usability, smoothing the relationship between product and user. Because investments in usability are less common, they make a wise strategy for differentiating one product from others. They also foster goodwill: Users feel that the company cares about their needs.

Marketing people routinely note the competitive advantage that comes with a user-friendly product. Ralph Thomas, formerly a group marketing manager at Ciba Corning Diagnostics Corp. of Medfield, MA, which designs and manufactures blood gas analyzers, is one who recognizes this advantage:

> If you place two blood gas analyzers in a laboratory, the lab technician naturally gravitates to the one that is easier to

use . . . These people often have too much work to do for the number of people available. They need a device that requires little training and little device preparation before running a blood sample.

He advocates designing products with high-quality user interfaces that make jobs easier, although he feels that companies should avoid overdesigning interfaces and missing their price points. He welcomes new design approaches that help to strike the right balance.

The first time that users try a product, enhanced user interfaces enable them to draw on their intuition to perform basic tasks, and this ability to perform the tasks on their own leads to a heady feeling of capability. When these new users take a liking to the product, they are apt to extrapolate the sense of quality afforded by the interface and project it onto other aspects of the product. Thomas agrees that there can be a halo effect and adds,

A good user interface says something positive about the company as well. It says that the company is talking to its customers, finding out what they want.

If initial ease of use gets a sales opportunity off to a good start, it makes sense for marketing to ask engineering to pay close attention to this attribute. Because many people view product usability as an unspecifiable intangible, what starts as an emphatic usability requirement often boils down to the familiar stock of platitudes, the so-called motherhood statements:

- The product must be intuitive to use.
- It must be easy to use.
- It must enable users to complete tasks quickly.
- It must minimize the frequency and consequences of user error.

In the end, such vague statements have little impact on user-interface design and product; usability becomes a crapshoot, producing wide variations in interface quality.

The fundamental problem is that most developers have not learned to think about designing for usability in a systematic way, in specific, quantified terms. They still believe that it is a matter of opinion, common sense, or consensus. You can usually tell how sophisticated a company's approach to usability is by asking marketing managers about their goals for a specific product. If they answer with a vague motherhood statement, such as "We want our products to be friendly for everyone," it is a good bet that they do not have a systematic, concrete approach. In contrast, the same people can be quite concrete about conventional engineering issues, citing engineering goals such as the following:

- Total product weight shall not exceed 4.5 pounds.

- The product shall not be damaged by a drop of 12 inches.

- Power consumption shall not exceed 10 watts.

- The control panel shall be impervious to noncorrosive fluids.

These goals are effective for three principal reasons. First, they are clear. Design team members can relate them to their own design challenges and can anticipate trade-offs. Second, they quantify performance, allowing comparisons with existing products and facilitating performance-level negotiations if trade-offs against other goals become necessary. Third, they permit product performance to be tested to see if goals are being met.

If usability goals are to be effective, they must have these same three characteristics. Fortunately, writing effective usability goals can be a relatively straightforward task (Bennett et al. 1988).

Writing Usability Goals

A good starting point for writing usability goals is looking at examples, such as those presented here for a hypothetical fetal heart monitor. Each goal focuses on a particular attribute of usability, such as speed of use; each is constructed around a specific metric, such as task speed; and each stipulates a performance level, such as 2 seconds.

Examples of quantified goals for a hypothetical fetal heart monitor are the following:

- On average, adjusting the heartbeat volume from minimum to maximum must take less than 2 seconds.

- On average, detecting the heart rate alarm must take less than 2 seconds.

- On average, replacing the recorder paper must take 70 seconds.

- At least 90 percent of users must correctly interpret the units of time printed on the recorder paper.

- At least 80 percent of first-time users must correctly attach the heartbeat sensor strap to the patient's abdomen.

- At least 80 percent of users must agree that the monitor cart is easy to maneuver.

- On average, users must be able to resolve the fetal heart rate with ±1 beat/minute precision.

- At least 90 percent of users must prefer the new product to the one it will replace in the current product line.

- At least 75 percent of users must prefer the new product to the [name of the leading product].

It can be tempting to base such a list on personal experience. For example, a project engineer who has worked on a certain type of product for several years might consider herself or himself the best judge of what user interface to include. Such an approach, however, can lead to products that are overcomplicated and that do not emphasize what is important to users. A better approach is to give potential users a voice in the goal-setting process through focus groups, structured interviews, and questionnaires. Taken together, these can give a more accurate understanding of how users are likely to interact with a given product, what aspects of the interaction are most important to them, and what level of product performance they find acceptable or unacceptable. Developers who thought they had a solid handle on users' needs come away from such an approach enlightened and far better prepared to write usability goals, a process that involves defining a set of usability attributes, writing the goals, and setting performance levels.

Defining a Set of Usability Attributes

The first step is to compile a list of usability attributes, based on input from users and on engineering judgment. Usability attributes are general aspects of user-product interaction. The list might include the following:

- First impression
- Long-term ease of use
- Initial ease of use
- Task speed

- Task accuracy
- Perceived complexity
- Error avoidance
- Recovery from errors
- User comfort
- Aesthetic appearance

- Task precision
- Sense of control
- Error detection
- Need for user manual
- User satisfaction
- Durable appearance

Obviously, this list is not exhaustive. To keep usability goals to a manageable number, however, the list of attributes should be pared to 15 or 20 of the most important.

Writing the Goals

The next step is to write at least one detailed goal for each attribute. For some attributes, such as task speed, one might write 20 or more goals based on the number of discrete user tasks. For an attribute such as aesthetic appearance, however, one might write just one goal. Each goal should use either an objective or a subjective measure. An objective measure is an observable aspect of user interaction, such as the time needed to complete a task or the number of errors committed while performing it—in other words, some aspect of interaction that can be timed or counted. A subjective measure is a user's opinion of the product (i.e., the individual's emotional response), expressed as a rating or ranking.

The goals presented earlier for a fetal heart monitor include both objective and subjective aspects of user interaction. For example, "On average, replacing the recorder paper must take 70 seconds" is an objective goal because task performance can be measured objectively; it is not affected by the user's opinion. In contrast, "At least 80 percent of users must agree that the monitor cart is easy to maneuver" is a subjective goal because it depends on the user's opinion. Opinions may be captured with a rating instrument in which a user is asked to agree or disagree with a statement on a continuum in which, say, a rating of 1 indicates strong disagreement, a rating of 5 indicates strong agreement, and a rating of 3 indicates a neutral response. To counteract users' tendency to take an overly positive view of even a difficult interaction with a product, surveying opinion with the scale in Figure 8.1 may be useful.

Figure 8.1. *Sample usability rating scale that clearly differentiates negative, neutral, and positive ratings.*

Poor						Good
-3	-2	-1	0	1	2	3

Setting Performance Levels

The most challenging part of writing usability goals is setting the performance level. How is a developer to decide whether users should be able to replace the recorder paper in 70 seconds or 100 seconds? Or what percentage of users should prefer the developer's new product to an existing one? Setting the performance level too high may place the goal out of reach for technical or human factors reasons, or it may require excessive time and money that might be better spent elsewhere. But setting the level too low may relegate the product to being an also-ran. Setting appropriate performance levels requires collecting data usability of comparable and competing products.

If developers are set on building the most usable fetal heart monitor, they must study the best-performing competing products as well as any products in their own line. In the total quality management vernacular, this effort would be regarded as benchmarking. One approach is to collect information by observing people as they use the products and recording information such as the time it takes them to perform a specific task. Depending on circumstances, the observer may watch without interfering or may cue the user. After the tasks are completed, the user may fill out rating or ranking forms designed to capture opinions.

Another approach is to conduct usability tests of relevant products (see chapter 23). Collecting data through tests involving about 10 subjects produces a reasonably reliable reference point for setting performance level goals. Many companies already operate competitive analysis labs and maintain an inventory of competing products; testing these products for usability is a natural extension of competitive analysis.

Developers should always collect usability data on their own products, particularly when performance level goals are being set for a new product that is intended to replace an old one. These data can ensure that the usability and the trouble spots of the existing product are understood well enough so that goals for an improved product can be formulated.

Usability research on the task of replacing the recorder paper of the fetal heart monitor might generate the data in Table 8.1. Such data facilitate setting the performance level in an informed manner. One might wonder, "Should we compete with (or exceed) the average task success rate of Product C?" Developers must know their own capabilities and constraints to answer this question. It may help to define three performance levels: the proprietary performance level is that attained by the developer's existing product (if there is one), the best-of-class performance level is the best among the competing products, and the breakthrough performance level is one that requires an accompanying breakthrough in user-interface design or technology.

Some developers might want to plug these performance levels into a formula to calculate a target value. The following formula, adapted from one used by the project review and evaluation technique, sets goals that counterbalance high hopes with realities.

$$\text{Performance level} =$$
$$0.2(\text{proprietary} + \text{breakthrough}) + 0.6(\text{best of class})$$

In the case of the hypothetical fetal heart monitor, the proprietary performance level was 85 seconds and the best-of-class level was 73 seconds. But suppose that the developer's engineers report a new concept that may enable users to change the recorder paper in 45 seconds or less—a major improvement,

Table 8.1. *Hypothetical data from usability research on the task of replacing the recorder paper of a fetal heart monitor.*

Product	Average Task Time (sec)	Success Rate (%)
A[a]	85	95
B	73	90
C	125	100

[a]Produced by the company developing the usability goals.

but one that will require a mechanical engineering break-through. Based on these values, the formula yields a target performance level of 70 seconds.

Implementing Usability Goals

To ensure design team commitment to usability goals, it is important to have design team members comment on them if not actually participate in their development. As with any performance specification, the team needs to recognize the importance of the goals and be prepared to accommodate the constraints they may place on product design.

Companies developing usability goals for the first time might do well to call a meeting—at least for key personnel—to describe the usability goal-setting process and emphasize management's commitment to it. When draft goals become available, they should be distributed to the key people in product development and marketing. If product design evolves in a direction that will not meet the goals, the deviation should be handled as any other deviation from a goal (specification) is handled. In short, usability should be addressed as rigorously as more traditional design concerns.

Usability tests on preliminary designs and on the final product indicate acceptability or unacceptability at each stage. By the time the final product has been tested, it should show significant usability advances that distinguish it from competing products. Usability then becomes a major selling point, and any questions about usability goals can be answered concretely.

References

Bennett, J., K. Butler, and J. Whiteside. 1988. *Usability engineering.* Tutorial presented at the Conference on Human Factors in Computing Systems, 16 May, in Washington, DC.

Chapter

9

Computer-Based Tools for Anthropometric Design

Modern aircraft cockpits are a testament to good anthropometric analysis—determining the best possible physical fit between people and their environment. A well-designed cockpit puts the important controls where the pilot(s) can reach them, provides a clear line of sight to instrumentation and the outside world, and provides a comfortable sitting position for people of varying sizes. This degree of harmony between humans and machines is the result of a comprehensive anthropometric analysis, which applies data on the size, weight, strength, and dynamics of the human body to design problems. Naturally, this form of harmony is of great interest to medical product designers, who must consider the physical requirements of medical workers and patients alike.

As it turns out, designers of medical products face many of the same anthropometric design challenges as cockpit designers, but they have not enjoyed the same benefits of computer-based analysis techniques that prevail in aerospace. To date, medical designers for the most part have used manual tools, probably because the cost of computerization has been too high for the benefits received. Now, a new generation of software tools stands to make computerization more appealing to medical designers.

Designers Assess Potential of Computer-Based Tools

Dale Foster is director of strategic product development for Hill-Rom, a manufacturer of hospital beds and wall systems in Batesville, IN. Presently, his industrial designers conduct anthropometric assessments of their products by measuring clearances on computer-aided-design (CAD) drawings, using data from a set of nomographic charts called *Humanscale* (MIT Press, Cambridge, MA). Foster thinks that integrating a computer-simulated human model with Hill-Rom's CAD approach could be the appropriate next step for the company.

> Our hospital beds have a lot of articulating components such as the sideguards and the mattress. We could use the tool early in the design process, for example, to make sure the sideguard controls are within the patient's reach.

Looking beyond the immediate benefits to product design, Foster feels that printouts of the computer-based analyses could be helpful in marketing the product.

> We could show our salespeople and customers how the bed dimensions match users' needs.

Christopher Goodrich, a senior industrial designer with Ohmeda, a manufacturer of anesthesia workstations in Madison, WI, says,

> A computer-based anthropometric analysis tool would be useful for exploring workstation design and iterating toward a final, detailed design.

Goodrich still sees a need for constructing a full-size mock-up of the detailed design, however, so that users can work with it and provide feedback to designers. He thinks that a computer tool could help detect physical interaction problems earlier in the design process, enabling his company to build mock-ups that are closer to the final design. Goodrich feels that

> interconnectivity between an anthropometric analysis application and the CAD system is key.

He is concerned about the applicability of existing anthropometric data to his design issues, however, pointing out that

> most of the available data are drawn from military personnel and may not be accurate for our user population.

He would like to see a research project aimed at building an anthropometric database for the medical worker population.

Potential Design Applications

What types of medical or diagnostic products would benefit most from rigorous anthropometric analysis assisted by software tools? An anesthesia workstation is one among many strong candidates, because there is substantial physical interaction between an anesthesiologist, his or her patient, and an anesthesia workstation. In a typical scenario, the anesthesiologist (or nurse anesthetist) may need to hold a mask over the patient's face with one hand while manually ventilating the patient using the rebreathing bag, adjusting the flow of anesthetic agent, setting alarm limits on the patient monitor, or making entries to the anesthesia record with the other hand. It is also necessary to make routine visual inspections of the breathing circuit, flow indicators, and vital signs display. During all this activity the anesthesiologist may be standing or seated.

An anthropometrically suitable anesthesia workstation should facilitate these activities from the intended use positions and should improve the clinician's productivity by shortening his or her motions. Furthermore, a workstation should provide a modicum of comfort over periods of extended use. After all, clinicians have no use for neck strain from repeated head turning or backaches that develop from overreaching. A complete study of reach and vision envelopes and lines of sight is necessary to achieve these goals and avoid these user problems (Van Cott and Kinkade 1972).

Other possible applications that could benefit from a rigorous anthropometric analysis include workstations employed in ultrasound, X-ray, magnetic resonance imaging, computed tomography scanning, mammography, and similar diagnostic procedures. Figure 9.1 shows how two human models, representing the patient and clinician, can be manipulated to assess the utility of a hypothetical workstation. Simpler medical products can also benefit. For example, an analysis of hand size and strength could be used as a guide for shaping and positioning hand grips or special controls on a device.

Overview of Anthropometric Data

Various databases contain values (usually mean and standard deviations) for stature, sitting height, knee height, buttock-to-knee length, seat breadth, weight, grip strength, and many other physical traits. Readily available sources contain data for

Figure 9.1. *Simulation of a hypothetical medical procedure produced with the Jack® software package.*

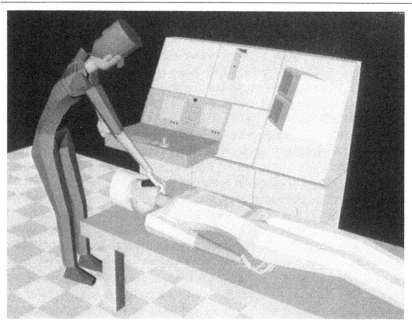

Created using Jack®, Center for Human Modeling and Simulation. Jack® is a registered trademark of the Trustees of the University of Pennsylvania.

subpopulations such as males, females, adults, children, the disabled, the elderly, and people of a particular occupational background and nationality. For example, a wealth of data on U.S. Air Force personnel has been collected for aircraft-design purposes. Because there is a lack of data on medical workers, however, medical product designers must use data from other populations.

Typically, anthropometric data are tabulated in terms of gender and percentile. For instance, one can look up the mean stature for the 5th-percentile female (60.0 inches) and the 95th-percentile male (73.9 inches) (NASA 1978). If the need arose, one could calculate the 47th-percentile stature from the statistical data. It is usually sufficient, however, to look up the extremes and perhaps the average (50th percentile).

Dimensional data fall into two general categories: structural and functional. Structural data describe standard body positions, such as a person standing symmetrically and erect. This provides a static view of the body that is useful for studying physical-clearance issues. Functional data describe real-life positions, such as bending forward and sideways at the same time to reach a control. They reflect the fact that the body is a complex set of parts that act together in ways that can extend a person's range of motion. Using functional data leads to a more realistic analysis than using structural data.

Anthropometric databases are built into computer-based anthropometric analysis tools. This relieves designers of the need to gather relevant data before moving forward with an analysis.

Software Tools

A recently released software package, Mannequin® (HUMAN-CAD, Melville, NY), provides a means for three-dimensional anthropometric analysis. The product was originally developed for in-house use by the ergonomists at Biomechanics Corporation of America (BCA) and since then has been refined into a commercial product. The PC-based software generates moving human likenesses that can be combined with images imported from several CAD and graphic design software packages such as AutoCAD, Corel Draw™, and Harvard Graphics. The software also has its own built-in graphics capability. Users select one of five body sizes, which range from "extra small" to "extra large," from pull-down menus or they can specify actual dimensions or percentiles. There are three body types to chose from: thin, average, and heavy. The human model can be rendered in three degrees of detail, ranging from a stick figure to a smooth-looking mannequin. The model is formed from about 3000 polygons that define the body's contours. Figure 9.2 shows the wire-frame model underlying the more finished-looking human models.

According to Dr. Clifford Gross, the product designer and CEO at BCA, this high fidelity is made possible by sonar maps of human beings and the development of algorithms to convert two-dimensional data on body segments into a three-dimensional model. Gross believes that

> Manneguin will bring convenient anthropometric analysis to the masses the way AutoCAD brought computer-aided-design to the masses.

To attain reasonable computer response times, Gross suggests using an IBM-compatible 386 computer (16 MHz, 4 MB RAM) as the minimum hardware platform, although the software will run on a 286 computer with a math coprocessor. The human model

Figure 9.2. *Wire-frame underlying the more realistic-appearing images produced with the Mannequin® software package.*

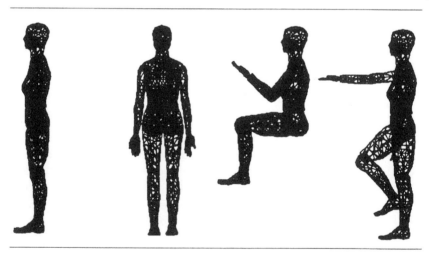

Produced with Mannequin® Designer. Mannequin® Designer is a product and trademark of HUMANCAD, a Division of Biomechanics Corporation of America in Melville, NY.

can be manipulated into realistic positions. Figure 9.3 shows the model's fully articulated hand, which can assume eight standard positions. Other interesting features of the application are the capability to display what the human model "sees" in its virtual world and to have it go for a walk.

Figure 9.3. *Computer model of the hand produced with the Mannequin® software package.*

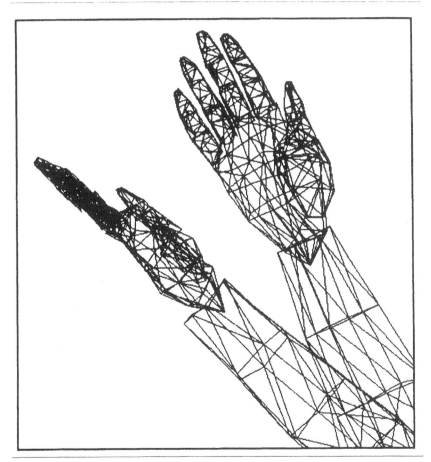

Produced with Mannequin® Designer. Mannequin® Designer is a product and trademark of HUMANCAD, a Division of Biomechanics Corporation of America in Melville, NY.

A product called Jack® (Center for Human Modeling and Simulation, University of Pennsylvania at Philadelphia) is at the higher end of available computer tools in terms of cost and hardware requirements. Jack® is a product of the University of Pennsylvania's Computer Graphics Research Lab, headed by Dr. Norman Badler, a recognized leader in the field of human modeling. Jack® runs independently on a Silicon Graphics IRIS 4D workstation and requires conversion software to export data to other CAD applications. The lab's staff normally provides a day of hands-on training to software purchasers and then is available for technical support. Users select a standard anthropometric model or specify the size of individual body parts via a spreadsheet.

Cary Phillips, chief systems developer for Jack®, says,

> One of its strengths is the ability to move body parts by clicking on them with a mouse and dragging them into the position of interest. Jack uses a general-purpose inverse kinematics algorithm to control the posture of the entire figure in a coordinated manner. The standard model has 88 degrees of freedom, including a fully flexible torso.

Jack® has been used by the farm, industrial, and outdoor power equipment manufacturer Deere & Company to study the position of bulldozer operators. According to Phillips, Jack® was set up so that the human model's hands maintained a grip on the controls while the rest of the body assumed various positions necessary to make visual checks on the bulldozer's operation. Jack® has also been used in the anthropometric analysis of the Boeing Apache helicopter's cockpit. Mike Prevost, a consultant to NASA's Ames Research Center (Moffit Field, Mountain View, CA), says,

> Jack answered critical design questions such as: What can the pilot see? How do the restraints constrict overall movement?

He says NASA's Ames is presently working to incorporate Jack® into simulations of pilot activities. Prevost estimates that it takes from two weeks to a month to become proficient with Jack®.

Alternatives to Computerization

Persons who lack access to software tools or who have no interest in using them can undertake a manual anthropometric analysis, although it is likely to be time consuming and tedious. First, there is the task of pulling together published data on the size and shape of people. Because of gaps in the various data sources, one might have to mix data from several populations to define a complete individual, which raises concerns about dimensional accuracy. Then, there is the task of accurately drawing people of varying sizes and postures, just as they would appear interacting with designs under development. Such drawings usually show people in standard positions, such as standing, seated upright, and reclined. Numerous drawings are required to investigate a complete range of body sizes and shapes. A shortcut is to draw only the anatomical landmarks of interest, such as the locations of the top of the head, eye location, and tip of the knee. This, however, can make it difficult to conceive exactly how people fit with a design—design flaws, such as insufficient clearance for a large abdomen, might be overlooked.

Anthropometric manikins are a compromise between the rudimentary analysis described above and analysis using software tools. Mannequins are flat, mechanical models consisting of hinged body segments. They can be constructed of clear plastic from published patterns or purchased from retailers (Church and Phillips 1985, 162–166). A rugged set of half-scale mannequins, including the 5th-, 50th-, and 95th-percentile male and 5th-percentile female, is available from Universal Energy Systems (Dayton, OH). Mannequins can be placed on top of a scale drawing and moved by hand into postures of interest. Typically, the limits of joint movement are marked to avoid overflexing or overextending a body segment.

Conclusion

Designers will continue to depend on real people to make final assessments of their designs. In fact, many companies have designated 5th-percentile female and 95th-percentile male employees to participate in design assessments. The aim of an anthropometric analysis is not to displace users from the design process, but rather to move designers more quickly toward viable design options in situations when evaluations involving real people are not feasible. Because of the availability of cost-effective computer tools, a comprehensive, up-front analysis is easier than ever before. As more medical and diagnostic device developers apply the tools in design, common problems such as out-of-reach controls and obstructed views of displays should become less common.

References

Anthropology Research Project, Webb Associates. 1978. *Anthropometric source book: Vol 1. Anthropometry for designers.* NASA Reference Publication 1024. Houston: Lyndon B. Johnson Space Center, National Aeronautics and Space Administration.

Church, R., and M. Phillips. 1985. The development of two-dimensional general population manikins for deriving workstation ergonomic requirements. In *Proceedings of the 29th Human Factors Society Annual Meeting.* Santa Monica, CA: Human Factors Society.

Van Cott, H., and R. Kinkade, eds. 1972. *Human engineering guide to equipment design.* Washington, DC: U.S. Government Printing Office.

Section

4

Designing the User Interface

Chapter

10

The Value of Human Factors Guidelines

Looking for guidance on the human factors engineering of medical products? Since 1956 the designer's source has been updated versions of *Human Engineering Criteria for Military Systems, Equipment, and Facilities*, MIL-STD-1472, the military's guide to human capabilities (such as static muscle strength) and suitable design solutions (such as the appropriate diameter for a radio control knob) (Department of Defense 1989). In the latest terminology this standard is the "mother of all human factors standards" by virtue of its broad scope, its significant impact on design practice (military and nonmilitary), and its spawning of similar documents.

One such document is *Human Factors Engineering Guidelines and Preferred Practices for the Design of Medical Devices*, originally designated AAMI HE-1988, published by the Association for the Advancement of Medical Instrumentation (AAMI) in 1988. Recently, AAMI released an updated and substantially expanded version of the document, designated ANSI/AAMI HE48-1993. Compared to MIL-STD-1472, the document is

As of early 1994, copies of ANSI/AAMI HE48-1993 could be acquired through the AAMI order department at $59 each ($39 for AAMI members).

somewhat more responsive to the specific design interests of medical product developers, although more limited in scope. Those who seek a "cookbook" for user-interface design are liable to be disappointed. Neither AAMI HE-1988 or the succeeding ANSI/AMMI HE48-1993 can prescribe a user-interface design solution. Similarly, the guideline document is not a particularly effective yardstick for evaluating the integrated performance of a given product. Nonetheless, it provides useful guidance on discrete design issues.

History of Development

The roots of AAMI HE-1988 can be traced to the 1960s and the work of the American National Standards Institute (ANSI) Z79 committee established to address the safety and performance of anesthesia equipment and components. Leslie Rendell-Baker, M.D., professor of anesthesiology at Loma Linda University (Loma Linda, CA), chaired the committee from 1962–1968. He says,

> A significant portion of [the committee's] work and of the resulting standard, ANSI Z79.8-1979, focused on human factors issues, such as where to place controls and displays to make working with the equipment simpler and safer. Many of [the committee members] saw the need for a more generic standard addressing the human factors of all types of medical products, not just anesthesia equipment.

When AAMI responded to this need in 1979, Rendell-Baker was installed as the AAMI Human Engineering Committee's user cochairman. He recalls with candor,

> In effect, we took the military standard, rubbed out the word *weapon*, put in the word *device*, and used it as a first draft. After the draft spent several years in gestation, we finally chose to publish it [AAMI HE-1988] in 1988, even though it was on the verge of being out of date, and immediately started work on a revision.

The committee saw sufficient commonality between the human factors requirements of military and medical applications to draw heavily on the military standard, even though doing so leaves AAMI vulnerable to criticism that adapting the military standard was an expedient approach at the cost of relevance to medical product design. Nevertheless, the guidelines represent a starting point for AAMI and a signal to the medical product industry that human factors is an important design consideration.

During the early 1990s the AAMI Human Engineering Committee worked on revising and expanding AAMI HE-1988. It convened twice a year for one-day meetings that usually followed conferences conducted by the medical industry and human factors organizations. This approach allowed it to draw clinicians, technologists, and usability specialists who typically attended these conferences. One of the committee's major goals was to infuse the already published document with guidance on the design of human-computer interfaces. Such guidance was in short supply in AAMI HE-1988, despite the major influx of computer-based products in the 1980s. It also sought to add guidance on the design of alarms and the auditory presentation of information.

Between meetings committee members—who represent medical product manufacturers, the human factors profession, and the government (FDA and the Nuclear Regulatory Commission)—refined the document's content and addressed public comments on the original publication. Frank Block, M.D., an anesthesiologist at The Ohio State University Hospital (Columbus, OH), served as the committee user cochairman, and Christopher Goodrich, an industrial designer with Ohmeda (Madison, WI), served as the industry cochairman. Rendell-Baker continued to serve the committee as a reviewer.

About the Resource

The stated purpose of AAMI HE-1988 is

> . . . to provide ergonomic and human factors engineering guidelines so that optimum user and patient safety, system safety and performance, and operator effectiveness will be reflected in medical device design. These guidelines are intended to promote effective work patterns, ensure personnel health and safety, and eliminate or minimize environmental stress, design-induced human errors, distractions, and complexities associated with the use of modern medical technology. However, adherence to these guidelines alone does NOT guarantee a successful HFE design; appropriate testing and evaluation are essential (ANSI/AAMI HE48-1993, 2).

In addition to some general recommendations, AAMI HE-1988 provides guidance on the design of controls, visual displays, audio signals, and consoles. Sample guidelines are presented below:

Display Resolution (Section 10.1.2, p. 60)
Cathode-ray-tube and flat-panel-matrix displays may be used for pictorial, graphic, and alphanumeric displays if the quality of information produced is adequate for the intended use of the device. When a display of complex shape is to be analyzed, the smallest detail of interest should subtend not less than 20 minutes of visual angle at the longest intended viewing distance.

Visual Alarm Indicators (Section 11.5, p. 68)
The use of flashing lights should, in general, be minimized. When used as listed below [according to a classification for high, medium, and low priority alarms], the flash rate should be within the range stated in table 13 . . . with an ON cycle of 20 to 60 percent. The indicator should be designed to that, once energized, the light will illuminate and stay on, even if the flasher device fails.

Momentary Silencing (Section 11.13.1, p. 70)
An audible high- or medium-priority signal may have a manually operated, temporary override mechanism that will silence it for a period of time—for example, 120 seconds. After the silencing period, the alarm should begin sounding again if the alarm condition persists or if the condition was temporarily corrected but has now returned. New alarm conditions that develop during the silencing period should initiate audible and visual signals. In this instance the use of intermittent repeating audible patterns is preferred over that of a continuous tone.

If momentary silencing is provided, the silencing should be visually indicated. Momentary silencing of an alarm should not affect the visual representation of the alarm and should not disable the alarm. A periodic audible indicator may also be used while the signal is silenced.

Notes:
1) Alarm design is extremely important to prevent nuisance alarms (which will make users turn off alarms) yet always provide important alarms.

2) Standards for specific devices may prescribe additional limitations on momentary silencing.

Consoles (Section 7.4, p. 29)

The effectiveness with which operators perform their tasks at consoles or instrument panels depends in part upon how well the equipment is designed to minimize parallax in viewing displays, allow ready manipulation of controls, and provide adequate space and support for the operator. Although no single console or instrument panel configuration is suitable for all applications, certain configurations are more effective than others. [Referenced figures], which are based on anthropometric data for individuals sitting or standing with erect posture, provide dimensional and other design criteria for various types of consoles . . .

About half of the newly released, 88-page document (which replaces the earlier 35-page document) is devoted to figures and tables that augment the guidelines; an excerpt is shown in Figure 10.1. Many of the tables and figures provide anthropometric data for use in properly sizing devices to suit the anatomical requirements of users. In the face of new software tools for performing anthropometric analyses (see chapter 9), the utility of these data may be limited to rudimentary applications in which the dynamics of human movements are not a critical concern. Additional figures and tables provide a plethora of data on designing controls, perhaps because the anesthesia community (clinicians and manufacturers) has historically been well represented on the committee and effective control design is critical to the safety and usability of anesthesia equipment.

As AAMI's Committee worked to infuse AAMI HE-1988 with guidance on computer-based interfaces, Rendell-Baker commented,

> The military came up with their own standards for computerized devices [ESD-TR-86-278] (Smith and Mosier 1986), and [the committee] has been working to boil that down to something suitable for the medical manufacturers.

In hindsight, the committee tapped a valuable source, considering the military's significant investment in developing guidelines to improve the usability of software user interfaces. The resulting new section on designing the human-computer interface is 15 pages long.

Figure 10.1. *Sample anthropometric data from ANSI/AAMI HE48-1993: horizontal and vertical visual fields.*

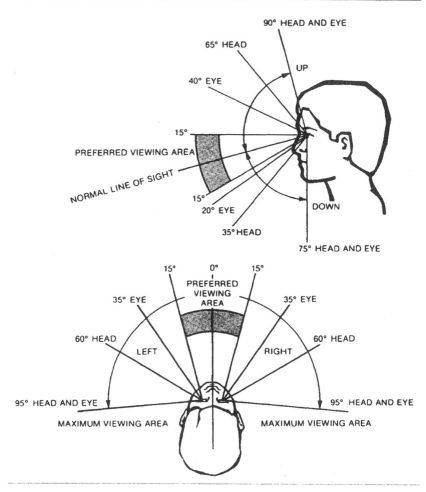

Courtesy of the Association for the Advancement of Medical Instrumentation: *Human Factors Engineering Guidelines and Preferred Practices for the Design of Medical Devices.* ANSI/AAMI HE48-1993.

Challenging the Usefulness of Guidelines

Around 1991, a research team, led by Richard Cook, M.D., and David Woods, Ph.D., of The Ohio State University's Cognitive

Systems Engineering Laboratory, challenged the usefulness of the AAMI guidance, particularly for the evaluation of microprocessor-based devices. Shortly after AAMI HE-1988 was published, the team set out to determine

> whether compliance with [AAMI guidelines] is in fact sufficient to produce designs with good human factors (Potter et al. 1990, 392–395).

To make this determination, the team used two methods to assess the human factors of an anesthesia circuit humidifier. The first method assessed the device in a static manner according to about 180 design issues cited in AAMI HE-1988. The assessment was conducted as if it were being done by an individual with limited prior experience in human engineering.

The second method assessed the device in a laboratory setting, enabling researchers to vary operating parameters and to introduce fault conditions; it statically assessed the device according to human-computer interaction guidelines drawn from prominent sources other than AAMI HE-1988. Human factors professionals were also involved in the assessment process.

Using the first method, the team found that the design violated 9 out of 60 applicable guidelines. They concluded, however, that

> with minor modification to displays and controls, [the device would] meet the AAMI compiled human factors recommendations. But meeting AAMI guidelines would not substantially change the user's ability to understand and operate the device (Potter et al. 1990, 392–395).

The latter part of this statement seems overly strong, considering that eliminating 9 deficiencies should improve the device's usability, whether or not it solves all the problems.

Using the second method, the researchers found design deficiencies that were not exposed using the first. For example, they found that because the device operates in two modes, warm-up and normal, users would likely have difficulty developing an effective mental model for device operation. Reportedly, the device does not provide users with a clear indication of its operational mode, thereby increasing the probability that they will become confused about its operational status. Clinicians invited to use and critique the device experienced the predicted confusion. The second method exposed additional problems, such as ambiguous alarm messages and inconsistent techniques for resetting the device.

According to the clinicians, these deficiencies, which have to do with design integration, are generally more problematic than discrete problems that might be detected using the AAMI guidelines. The researchers assert:

> These deficiencies arise because the intrinsic flexibility of microprocessor displays and controls frees designers from the limitations of older systems, where the relationships between displays, controls, and internal function were highly constrained. These deficiencies may be found in a large number of consumer devices (e.g., videocassette recorders), where they represent an annoyance, and also in more critical domains, where performance of the human operator is critical to safety or security (Potter et al. 1990, 225).

The thrust of the research discussed above is that guidelines, taken alone, are insufficient to produce a high-quality user interface. The researchers state:

> Detecting and correcting these deficiencies in device design requires knowledgeable application of the principles of human factors engineering rather than just more detailed or extensive guidelines (Potter et al. 1990, 392–395).

Their point is that human factors design is not a cookbook design, underscoring a concern that persons not fully versed in human factors engineering methods might apply the guidelines to a product in development, detect and fix a set of superficial problems, and then deem the product "user-friendly" and ready for market. Meanwhile, major problems having to do with the interactions between the device and its users might go undetected and lead to serious usability problems. The researchers state:

> The new generation of high-integration monitoring equipment is potentially prone to just [these sorts of problems] as multiple, discrete, hardwired devices are incorporated into a single shell under the aegis of supervisory control. . . . [I]ncidents are seldom single-point failures, but rather represent the confluence of multiple events, each alone insufficient to cause an incident but in combination leading to disaster (Cook et al. 1991, 225),

Conclusion

Is a set of imperfect guidelines better than no guidelines? In the case of AAMI HE-1988 and its enhanced successor ANSI/AAMI HE48-1993, the answer is yes, considering just the symbolic value of the publication—an endorsement of human factors knowledge and methods by the medical industry. Further, there is no question that the document contains valuable information, albeit largely similar to information contained in other sources, most notably MIL-STD-1472.

Yet, the Ohio State researchers make a strong point that the guidelines are not fully effective at identifying problems relating to the cognitive processes associated with using a product or with issues having to do with user-interface integration, a point supported by AAMI user cochairman Block. However, the purpose of this criticism is not to negate the significance of the guidelines but rather to point out the importance of involving human factors professionals in the design process and evaluating product designs in a more dynamic fashion (see chapter 23). Furthermore, the fact that researchers used the guidelines to detect nine deficiencies in an anesthesia circuit humidifier validates the utility of the guidelines.

Ironically, the most important target audience for the AAMI guidelines may not be product designers or users at all. Rather, it may be those in medical product marketing who can be influenced by the standard. Reflecting on the state of anesthesia equipment design, Rendell-Baker says:

> Better designs are held back by the marketing people who fear bringing progressive products to a conservative marketplace. The typical anesthesiologist, someone who has worked in the field for years, can be extremely conservative and look for new products that function much like existing products.
>
> If it [an anesthesia device] doesn't look like last year's machine, they don't like it. However, marketing people must understand that the major gains in human factors will come from advanced electronic designs that put information and control within easy access of the user.

Rendell-Baker believes that marketing people should show courage and bring progressive designs to market so that the transition to better, human-factored designs can occur sooner, rather than later. The next release of the AAMI guidelines should fuel this progress.

References

Block, F. 1991. AAMI human engineering committee. *STA Interface* 25(4):8.

Cook, R., S. Potter, D. Woods, et al. 1991. Evaluating the human engineering of microprocessor-controlled operating room devices. *J of Clinical Monitoring* 7(3):225.

Human engineering criteria for military systems, equipment, and facilities, MIL-STD-1472 (D version). 1989. Washington, DC: Department of Defense.

Human factors engineering guidelines and preferred practices for the design of medical devices, AAMI HE-1988. 1988. Arlington, VA: AAMI.

Potter, S., R. Cook, D. Woods, et al. 1990. The role of human factors guidelines in designing usable systems: A case study of operating room equipment. In *Proceedings of the Human Factors Society 34th Annual Meeting*. Santa Monica, CA: Human Factors Society.

Smith, S., and J. Mosier. 1986. *Guidelines for designing user interface software*, ESD-TR-86-278. Bedford, MA: Mitre Corp.

Chapter
11

Improving the Visual Design of Computer Screens

Many medical device companies are asking the same question: How can we improve the usability and appearance of the screens for our computer-based products? Some compare their own screen designs with the elegant designs of commercial software applications such as Microsoft® Word or Lotus 1-2-3® for Windows™, and feel they could be doing much better.

Device manufacturers may not realize, however, that commercial software developers invest considerable time and money into the quality of their screen designs, employing large staffs of user-interface design specialists who are committed to improving the screens' information content, flow, and appearance. Their strong commitment to quality screen design is necessary in the commercial software industry because consumers have become accustomed to good design and will no longer accept inferior solutions.

Device manufacturers may need to make the same commitment within the next year or two. After all, medical workers

125

are increasingly making use of commercial software applications in their jobs, and are coming to expect better design solutions from device manufacturers than they have been getting. As has been true for commercial software developers, device manufacturers may find that improving the screen designs of their user interfaces can make their products more competitive.

Standard User Interfaces

In the last few years the expedient solution for many device companies that have wanted to improve the appearance of their screens has been to adopt a standard, commercial user-interface style, such as OSF/Motif®, Microsoft® Windows, or OPEN LOOK. Developer kits and user-interface style guides are available for these and other graphical user-interface designs. Using these resources, it is a relatively simple matter for even inexperienced designers to match the look of a modern user interface. In fact, many of the development tools force designers to adopt a standard screen appearance that provides a modicum of consistency, readability, and attractiveness. Furthermore, adopting a standard user interface also facilitates coding, which may lower software development costs.

Let us focus for a moment on the visual design benefits of adopting a standard user-interface style. The leading commercial examples all look fairly elegant (Figure 11.1). Differences in style are functionally insignificant except insofar as they reflect differences in display resolution or the use of a color display. In fact, most interfaces mimic the original Apple® Macintosh™ user interface, which set the high-water mark for usability and visual appeal.

Following a particular style, development tools relieve developers of the task of designing user-interface elements—the so-called graphic user-interface widgets, such as dialogue boxes, windows, menu bars, control panels, and so on. Typically, the code to generate them is provided in the developer kits, enabling someone who has not been steeped in user-interface principles and graphic design to produce attractive, albeit standard results.

In certain cases prototyping is also easier when designers adopt a standard interface style. For example, an application called Visual Basic™ enables designers to model Windows-compatible user interfaces relatively quickly.

Perhaps the biggest payoff for adopting a standard graphical interface style is that users may already be familiar with other applications that share the same style and interaction

Figure 11.1. *Sample windows from OSF/Motif®, Microsoft® Windows, and OPEN LOOK.*

OSF/Motif

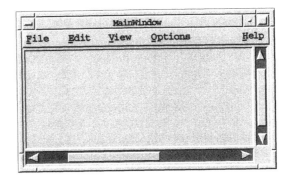

Microsoft Windows 3.0

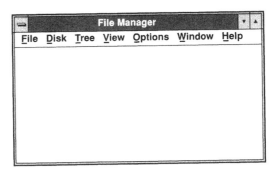

OS/2 Presentation Manager

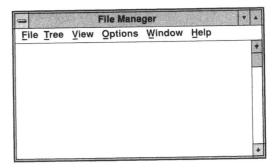

Courtesy ACM Press and Aaron Marcus + Associates, Inc.

methods, thereby making it easier to use a compatible product initially. This strategy worked well for Apple Computer beginning in the 1980s, when they and third-party vendors produced software products with compatible user interfaces while DOS users struggled to learn one novel application after another. Because standard graphical user interfaces are likely to dominate the commercial software industry for the balance of the 1990s, the temptation to adopt one is that much stronger.

But is it valid to assume that device users will start further ahead on the learning curve if the medical facilities in which they work use computer terminals or devices that run OSF/ Motif®-style applications? In medical environments, where a new user's computer experience may be limited to using a word processor in college, the real payoff for the manufacturer may come only when it brings additional products to market that employ the same interface style.

Manufacturers should also keep in mind that designing interfaces according to a "cookbook" solves only part of any screen design challenge, because a derivative of the "garbage in, garbage out" rule still applies. In other words, designers can make information look terrific, replete with three-dimensional effects, but it still may not serve users' needs. This is why it is important for manufacturers to follow a usability engineering process that involves several user research techniques, such as task analysis, to determine the information users need to accomplish their work.

Another shortcoming of adopting a standard user interface is that it may ultimately be a force fit for a given application. For example, most development tools work best on applications with a color screen, in which designers work with a palette of grays to creative effective three-dimensional effects that do not overwhelm the imbedded information. By comparison, most do not work well with a monochrome display that creates the effect of the color gray with a pattern of closely spaced dots. In such monochromatic applications, designers can inadvertently produce visual chaos or "patternitis" by placing too many patterns of dots in close proximity. Also, the readability of text is degraded when it is displayed on top of such a pattern (Figure 11.2).

Furthermore, how well a standard graphical user interface works depends on the size of the display. Many medical devices incorporate small displays with odd aspect ratios (different than the standard three units high by four units wide) that leave little room to incorporate the normal features of a standard graphical

Figure 11.2. *Text becomes increasingly harder to read as the background becomes darker (as the dot density increases).*

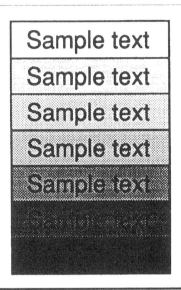

user interface, such as a window title area and a system-level menu bar, along with core information. In such cases, standard user-interface elements may ultimately confuse users, whereas a product-specific solution may better fit users' expectations. Of course, some user-interface designs overcome this problem by allowing designers to conceal certain mechanical elements of the development or windowing environment. Nonetheless, many manufacturers decide to develop their own user interface, using programming languages such as C++. This approach gives designers more creative control, but it also creates a risk that a visually inferior product will be developed if the right professionals are not included in the design process.

A final concern about using standard graphical user interfaces is that they often require hardware with substantial processing speed, memory capacity, and a pointing device (typically a trackball or mouse). This may not be a problem for products such as expensive diagnostic scanners or chemistry analyzers, which require large processing capacities regardless of user-interface style and are configured as workstations with space for a pointing device. However, hardware capacity may be the deciding factor for the interface design of a smaller and

relatively inexpensive product, such as an infusion pump, glucose meter, or even a patient monitor. Also, many such devices
cannot easily accommodate a pointing device.

Custom User Interfaces

The decision to develop a custom or product-specific user interface increases the screen designer's workload but, at the same
time, increases creative freedom, which can work in the favor of
development staffs that have the requisite talent. After all of the
necessary up-front work has been completed to define an effective cognitive model for the user interface, designers begin the
visual design process by creating an interface style guide and
so-called templates. In effect, they create the guidance materials that are normally provided with standard graphical user
interfaces.

Guidelines for screen design can be found in numerous
human factors and graphic design textbooks, and may also be
adapted from guidelines in commercial style guides, which are
typically available in retail bookstores and software stores.
Sample guidelines for typical, full-screen software applications
might include the following:

- Center the name of the application [for multiple application
  systems] in the middle of the title bar, at the top of the window.

- If the list does not fill the on-screen area allocated to it, do not
  display an elevator in the vertical scroll bar.

- Use a blue background color for all displays (Hix and
  Hartson 1993, 25).

For a smaller medical device incorporating a lower-resolu-
tion display, some useful guidelines might include the following:

- Indicate a selected menu option using inverse video.

- Place all status messages in a rectangular box with a 3:4
  aspect ratio.

- Separate the title field from the information field using a two-
  pixel-width line.

Templates are a visual presentation of much the same kind
of guidance contained in written guidelines. A given device
may require a half-dozen basic templates that dictate the screen

layout and the way the information appears. Ideally, the number of basic templates should be minimized for the sake of simplicity, while ensuring that screen layouts make it easy to distinguish and acquire different types of information. For example, a spreadsheet-type template can be used for listing patient data over time, while an on-screen control panel template might be used for device setup. A simple template is shown in Figure 11.3.

If designers approach the task of writing guidelines and developing templates with sufficient rigor, the balance of the screen design effort should progress smoothly. Then, design issues will center on the information content and flow, rather than on issues of appearance.

Design Considerations

Important design considerations for graphical user interfaces include screen size, information resolution, emergent features and analogs, and the use of color (see Table 11.1).

Figure 11.3. *Sample screen template.*

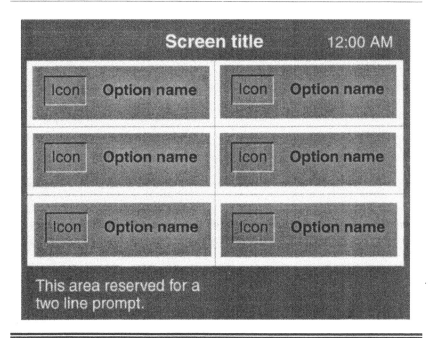

Table 11.1. *Checklist for effective screen design.*

- Check consistency of related visual elements.
- Limit highlighting.
- Avoid clashing visual elements.
- Use alignment grids when laying out information.
- Balance areas of information with white (blank) space in order to avoid screen vacancy or congestion.
- Use no more than five colors.
- Use colors according to a meaningful coding scheme.
- Limit the number of line widths (i.e., pen sizes)
- Limit the number of fonts and sizes.
- Justify left margins of text.
- Avoid arbitrary capitalization.
- Keep text short (40–60 characters) for legibility.
- Use fill areas in place of rectangles with distracting border lines.
- Proportion objects so that they are pleasing to the eye (rule: 1 unit wide, 1.618 units high).
- Use high contrast between figure and ground for readability.
- Use consistent spatial cues (layering, beveled edges, drop shadows, sizing, texture) for three-dimensional applications, and create three-dimensional effects with changes in the value (brightness) of colors.
- Base icons on common elements (e.g., always use scissors to symbolize cutting).
- Prefer graphics to words where they communicate more information in less space.
- Seek ways to simplify everything.

Adapted in part from guidance presented in:

Marcus, A. 1992. *Graphic design for electronic documents and user interfaces.* New York: ACM Press.

Tufte, E. 1989. *Visual design of the user interface.* Cheshire, CN: Graphics Press.

Screen Size

In the development of many medical devices, choosing a screen size is a critical and sometimes contentious matter. On the one hand, some divisions within the company may want to minimize screen size in order to maximize a device's portability. On the other hand, users may have specified a large body of information they need to have presented when the product is in its baseline, top-level, or "resting" state, which would require a larger display. A desire to minimize the number of layers in a menu system is another reason to incorporate a larger display.

The economics of display technology may introduce a cost penalty if designers select something other than the most common display size (e.g., a 14-in. color monitor or a 2 × 16 character-based LCD). Clearly, a 19-in. monitor costs more than a 14-in. monitor. However, because of economies of scale, 14-in. monitors commonly used in desktop applications are often less expensive than 9-in. monitors. Ultimately, developers may need to rely on a decision matrix or an executive decision to settle on the right size.

During the decision-making process, designers and decision makers should keep in mind that screen size has a dramatic effect on the user interface as a whole and the appearance of screens in particular. Generally, larger screens allow for better screen designs. The added space gives designers the freedom to place information where it belongs, rather than where it fits, and to compose uncluttered screen layouts from which users can obtain information quickly and easily. A rule of thumb on screen size is as follows: Within limits, the appearance and usability of screens improves as screen area increases. Figure 11.4 illustrates this general principle. Doubling the screen area, for example, might improve screen appearance and cut task-performance times substantially for certain applications.

Information Resolution

Most screen designs include extraneous or unrefined details and could use good editing to improve their information resolution. Edward Tufte, a graphic design educator at Yale University, explains:

> The two strategies for improving information resolution are (1) to reduce the noise and (2) to improve the signal.

In others words, designers need to ensure that the most important information stands out from the background elements, a concept familiar to electrical engineers but less familiar to screen designers.

Figure 11.4. *A small screen (a) may look congested and require users to access several screens to acquire the necessary information. A larger screen (b) provides space to lay out information in a pleasing, readable format. A particularly large screen (c) may overwhelm users and require them to spend excess time scanning for information.*

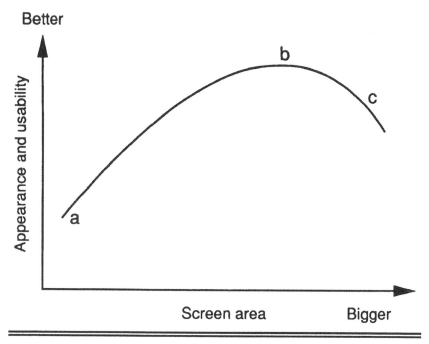

In a tutorial pamphlet prepared for the IBM design program, Tufte states:

> Noise reduction includes eliminating every pixel not serving user-important data, muting minor visual elements, removing distracting patterns, ending clutter (particularly that resulting from computer administrative debris), avoiding fussy redundancy in visual signals, and continually directing screen design toward a clean, sharp, precise user interface.

> Signal improvement includes turning as much of the screen real estate as possible over to user information, promoting widespread distribution of improved display technology, designing typography to classical standards, and minimizing jaggies in type and graphics through antialiasing techniques (Tufte 1989, 3).

Figure 11.5 contrasts low- and high-resolution images for a hypothetical medical device.

It is important to remember that increasing the information resolution does not necessarily mean increasing the amount of information presented in a given screen area. Designers should aim for a comfortable balance between the amount of white or blank space on the screen and the amount of information in the form of text, icons, and other features, so that users can acquire information quickly.

Emergent Features and Analogs

Some product developers and medical practitioners believe that in the future more medical device displays will incorporate so-called emergent features and analogs. An emergent feature is a meaningful by-product of two or more discrete visual elements (Bennet, Toms, and Woods 1993, 73). For example, on a graph, the area under a curve is an emergent feature that is the by-product of two axes and a plotted line. Emergent features can accelerate the acquisition of information or provide new information. An example of an analog display that would replace or supplement a digital one is using a pictorial representation of the patient's body to convey vital signs and an integrated sense for the patient's well-being, rather than simply presenting a matrix of digital values. In such a pictorial, the color of blood moving across the lungs would change from blue to red as it picks up oxygen, enabling clinicians to develop an "at a glance" sense of the patient's respiratory and pulmonary state.

Proponents of such displays suggest that they can reduce the mental effort required to extract information for immediate

Figure 11.5. *Comparison of a low-resolution and a high-resolution graphic element.*

decision making. However, some practitioners oppose such displays because they say that while the displays may make it easier for designers to understand what is going on with the patient, they are not helpful to clinicians who are accustomed to acquiring raw data about a patient's condition, then performing the necessary mental integration themselves.

Use of Color

Many companies have difficulty deciding whether to specify a color or a monochrome display. Of course, costs can be a major factor that designers must consider when developing products for price-sensitive markets—the price of a color monitor may be several times the cost of a monochrome one. Some manufacturers resolve the issue by offering either a monochrome or a color display, leaving the choice to users and their equipment-procurement budgets. However, this approach complicates screen design.

When designing monochrome screens, designers must use lines and patterns to segregate and code information. In the process they must be careful not to use too many patterns, since the interaction between patterns can produce unwanted effects, such as "patternitis." Therefore, designers should limit the use of patterns to one or two and rely more on the use of lines for demarcation. Unfortunately, developing a quality color design from a monochrome one is not a simple matter of coloring in fill regions and changing the color of lines—that is, taking a paint-by-number approach. This approach produces amateurish-looking screens. High-quality color screen designs require a more exhaustive remodeling effort. For example, upgrading to a color screen enables designers to do a good job of making the surface of the screen appear three-dimensional or, more specifically, $2^1/_2$-dimensional. The term $2^1/_2$-*dimensional* is used by some designers to describe user interfaces that appear to have a foreground, middle ground, and background within a closed space. Such screen depths appear to be about $3/_{16}$ in. thick, and are based on such features as a $1/_{16}$-in. beveled edge on the side of a dialogue box.

The key is using gradients and discrete gray tones to highlight and deemphasize surfaces, and bevels and drop shadows to create the sense of depth or etching. Some designers refer to the resulting set of gray objects as "grayware." Filling a rectangle with color does not achieve the same effect, nor does using varying density dot patterns on monochrome monitors.

More-elegant $2^1/_2$-dimensional user interfaces use colors other than gray only for accent. As with hot spices in foods, colors should be used sparingly and for important purposes only. For example, a list of digital values can be shown as black text on a light gray middleground, while a value for an alarm condition can be shown as bold white letters on a medium-red foreground. To avoid overuse of color, it may be best to include appropriate uses for color in a style guide. Color should be selected according to medical industry coding conventions, where applicable, and according to a simple, harmonious palette that is relaxing to look at (see chapter 12).

Most users prefer a good color screen to a good monochrome one. Color makes it easier for people to acquire and interpret information, because it can be used effectively to make important information stand out and as a coding technique to convey additional information, such as an alarm status (e.g., using red, orange, and yellow to convey three priority levels). Furthermore, as medical workers become more accustomed to interacting with color screens in nondevice applications (e.g., computers), they will come to expect them in medical devices. Therefore, using color as a display option is a prudent design approach for many products.

Who Takes Responsibility?

For certain applications the distribution of responsibility for design decisions is relatively clear. For example, in the course of erecting a skyscraper, architects focus on its overall shape, appearance, and internal organization; civil engineers work on the foundation and structural frame; and contractors place the concrete, steel, and glass where it belongs. Such a division of labor results from centuries of experience erecting buildings—a natural segregation of responsibility based on professional knowledge and skill. This approach contrasts sharply with the way most user interfaces for medical devices are designed today, display screens in particular.

The screens incorporated in many medical devices are created by people who lack the requisite training and experience to do a good job, despite their most diligent efforts. In some cases the surrogate designer is a software programmer who may be excellent at coding the user interface but, by his or her own admission, knows little about determining screen content and flow. Or, different parts of the user interface may be designed by people from varying disciplines who make up the design team. When individuals who do not have formal training in screen design take a crack at it, the results may look amateurish because of their odd layouts, ineffective use of color, ambiguous icons, use of jargon within text segments, missing elements (prompts, screen titles, screen navigation controls), and other common problems. This is not the outcome that quality-conscious companies seek.

Therefore, manufacturers need to seriously consider who is going to do their screen design work. Options include formally training the staff in graphic design principles through in-house workshops, professional seminars, and college courses, and then giving them time to practice designing screens that are consistent with a standard graphical user-interface style guide. This is a long-term, "bootstrapping" approach that may or not pay off, depending on the talent of the people involved. A more expedient option is to hire or retain the services of user-interface design specialists who have the skills and experience to take a basic design and produce good-looking screens. A good source for such persons might be designers working for commercial software developers who are looking to change jobs or have fallen victim to the ongoing industry shakeout. Manufacturers might also retain the services of freelance graphic artists who specialize in screen design. Contact the National Computer Graphics Association (Fairfax, VA) or the Association for Computing Machinery's Special Interest Group on Graphics and Interaction (New York City) for more information on this subject.

Perhaps the best long-term solution for larger companies is to create a user-interface design group that can meet screen design needs as well as a broader set of usability engineering needs. A key member of such a group would be a graphic artist. Such individuals can work closely with usability engineers to make sure that screens incorporate usability principles, but also exhibit an artistic quality that will increase product appeal and marketability.

Conclusion

It is often thought that screen design is a superficial aspect of a medical product, and that medical workers only need devices that "get the job done." But commercial software products demonstrate that good screen design is linked to the usability of the product. Therefore, good screen design is an important factor in product differentiation. Accordingly, medical device manufacturers should place screen design high on their design-needs agenda, showing greater sympathy toward users who may spend hours a day looking at a given display over its life cycle of several years. They should continually ask themselves, What kind of display would I want to look at for 300 hours a year over the next 6 years?

Years ago, few people knew how to design good-looking, functional screens. Fortunately, the crop of designers capable of creating modern-looking software has grown substantially. Medical device companies should carefully consider the benefits of hiring or retaining such persons to improve the quality of their products. It is a small price to pay to improve a characteristic that has such a large bearing on customers' initial and lasting impressions of a product as well as its usability.

References

Bennet, K., M. Toms, and D. Woods. 1993. Emergent features and graphical elements: Designing more effective configural displays. *Human Factors* March:73.

Hix, D., and R. Hartson. 1993. *Developing user interfaces—Ensuring usability through product and process.* New York: John Wiley.

Tufte, E. 1989. *Visual design of the user interface.* Cheshire, CN: Graphics Press.

Chapter

12

Making Color a Contributing Component of the User Interface

The effective use of color can improve the usability of most medical products that involve user interaction. However, to use color effectively, designers first must understand the principles of color and then apply them in a systematic manner appropriate to the product being designed, the people who will use it, and the use environment. The systematic application of color can elevate the role of color in product design from mere decoration to that of a contributing component of the user interface. Human factors experiments show that color can enhance an individual's performance of tasks, such as visual searching and tracking. Therefore, designers should assume that color has a significant effect on a product's overall usability and, for that reason, should consider color carefully in the design process.

However, many human factors specialists and graphic designers have found that working with color is especially challenging; good-quality designs do not come easy. For example, color can be intimidating and hard to work with because its effectiveness is measured both objectively and subjectively. Physics and human factors data tell us which colors are appropriate to fulfill design objectives (such as rapid detection) and are likely to produce a good appearance. However, because color affects people differently on a subjective basis, a specific use of color(s) will disturb some users. Ultimately, the designer's experience, sense of good color design, and assessments of existing products all must be factored into the design process, augmenting the objective criteria.

Color is used extensively in common medical products, from advanced electronics to disposable plastics. It serves various purposes, including drawing the user's attention to parts of the user interface and communicating specific messages. For example, coloring the start key green on a pump control panel will make it stand out against an array of 20 keys of other colors (Figure 12.1). Such highlighting makes sense particularly if the key is used frequently or if immediate access to the key is important. Similarly, highlighting with color can help a user either find a specific numeric value (such as body temperature) in a column of values on a computer display or select the correct ECG lead among a cluster of leads. Before deciding which areas should be highlighted on any given device, designers should

Figure 12.1. *Color can be used to draw the user's attention to a control. (Note: Black is used to simulate a colored key in a group of same-colored keys.)*

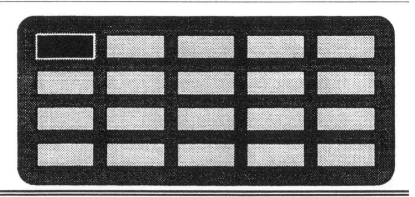

analyze the tasks that users perform and the flow of information between the user and the device.

Color also helps the eye track a moving image. This factor becomes evident if one examines a moving wave form (such as a blood-pressure trace) on a monochrome monitor and then on a colored patient monitor. On the monochrome display, irregular lines that overlap repeatedly can prevent the user from discriminating individual traces. In contrast, color-coded traces are easier to discriminate from one another (Figure 12.2). In a recent study the majority of physicians preferred color over monochrome traces (Wiklund, unpublished).

Figure 12.2. *Black-and-white waveforms are difficult to discrminate when they are spaced closely together. Color can help users discrimate individual waveforms.*

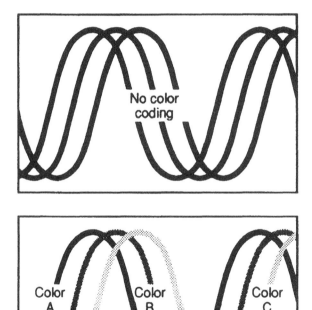

Color by itself can communicate information, as is demonstrated by the everyday example of traffic lights. Alarm systems communicate information in a similar manner for a variety of other equipment, using red, orange, and yellow indicators to convey varying levels of threat or hazard. In fact, the American Society for Testing and Materials (ASTM) has formed Subcommittee F29.03.04 to consider standards for the harmonization of alarms with implications for color selection for warning information.

Designers can also reinforce functional groupings by using the same color (which may appear to be a shade of gray) for elements of controls and displays. Of particular importance to marketing personnel is the fact that color can enhance the aesthetics of a product and should, therefore, not be ignored, because purchasing decisions are indeed influenced by design aesthetics.

A more fundamental issue regarding the use of color is how color selection affects information recognition, discrimination, and legibility. Research into the physics of color and human perception of color tells us that, to maximize legibility, the contrast ratio between a colored object and its background should be at least 7:1; that white, green, and yellow characters are most legible on a black background; and that black and blue characters are most legible on a white background (Thorell and Smith 1990). Figure 12.3 shows how text legibility varies as the background color shifts in hue and value. Reviewing principles of

---

Figure 12.3. *Text legibility changes as value and hue (simulated with grays) change.*

color use such as these can help designers avoid fundamental errors in color selection.

At the same time, designers should consider potential use environments for their products, because lighting conditions can have a significant effect on how people perceive color. Maintaining strong differences in value (the brightness of a color measured over a range of black to white) between the color of objects and their backgrounds helps to maintain legibility in low lighting conditions. It also helps individuals whose color vision is impaired.

> The most common forms of color vision deficiencies are the inabilities to discriminate red from green and blue from yellow (Hurvich 1981; Smith 1988).

Furthermore, 8 percent of males have difficulty in distinguishing between pale pastel shades of green, red, gray, and beige. Combinations of these colors should, therefore, be avoided.

The effectiveness of color on human performance has been researched extensively. A particularly conclusive study (Thorell and Smith 1990) is discussed in Thorell and Smith's authoritative book, *Using Computer Color Effectively.* In a study involving air traffic controllers at Gatwick Airport in the United Kingdom,

> test subjects were trained to locate planes on either a color or monochrome display. Those trained with the color displays attempted to locate distinctly different colored targets; the other group tried to locate targets highlighted by other methods. All other aspects of the displays (for example, screen resolution and software) were the same. The test results showed that subjects using color displays located aircraft 16% faster than those using monochrome screens.

Even though this research addressed the interests of air traffic control system designers, the findings have applications in medical product design as well. Many more studies on the effectiveness of color provide equally solid evidence that color can improve human performance. Therefore, designers should make it a short-term initiative to search available literature for color-design information. However, a review of the literature may still leave designers feeling that a shortage of guidance exists regarding actual color selection. Without efforts to give designers more guidance, color selections are likely to be made at the individual design-team level, thus forestalling product class integration.

Given the present lack of convergence toward standard uses of colors within most product classes, medical workers find themselves working with product after product that speaks its own visual language. Consequently, users no longer associate a meaning with a particular color. If used conservatively, colors may still do a good job of drawing attention to areas such as a specific control, but the additional meaning that color is intended to communicate is lost. Information embedded in a color code fails to reach the user and, therefore, users must continue the task at hand without the benefit of the information.

Aaron Marcus is a recognized authority on color graphics as applied to product design and communication media and is the author of *Graphic Design for Electronic Documents and User Interfaces* (ACM Press 1992).

> One typical mistake is to use too many saturated colors that end up competing for the viewer's attention—the so-called Las Vegas approach to user interface design. People like to visit Las Vegas but they don't necessarily want to live there. Another typical flaw is a lack of hierarchy; not being sensitive to the hierarchies that color can establish; not respecting the nuances of relationships between parts of the user interface; having too many unorganized colors that fail to convey meaning.

To solve the problem of indiscriminate or overuse of color, Marcus recommends

> solving the design problem in black and white first and then adding color according to an established hierarchy.

In one article Marcus says,

> The basic idea is to use color to enhance black-and-white information (Marcus 1990, 137).

He refers designers to *Interaction of Color,* a basic text that will help designers develop appropriate color palettes for products (Albers 1975). In another article Marcus defines what he calls "The Ten Commandments of Color":

1. Use a maximum of five, plus or minus, two colors.

2. Use foveal (center) and peripheral colors appropriately.

3. Use colors that exhibit a minimum shift in color/size if the colors change in size in the imagery.

4. Do not use high-chroma, spectrally extreme colors simultaneously.

5. Use familiar, consistent color-coding with appropriate references.

6. Use the same color for grouping related elements.

7. Use the same color for training, testing, application, and publication.

8. Use high-value, high-saturation colors to draw attention.

9. If possible, use redundant coding of shape as well as color.

10. Use color to enhance black-and-white information (Marcus 1986, 7–11).

A Designer's Viewpoint

Larry Hoffman, an industrial designer with Hewlett-Packard's Waltham Division (Waltham, MA), participated in the development of his company's new component monitoring system, known less formally as "Merlin."

> I would like to be able to turn to a color standard to help make certain design decisions, but there is not much out there that applies directly to medical products. Designers from other divisions within Hewlett-Packard frequently call me to find out if I know of any standards for the use of color on a specific type of product. We try to look at other products on the market to see where we can be consistent, but often we just have to make our own decisions about color. We also need to work within a limited palette of HP colors that serve to give HP products a uniform look.
>
> In my designs, I try to keep the high-contrast, high-visibility stuff [colors] up front where it does the most good. I try to accentuate the user interface and make the rest of the product disappear into the background where it does not draw attention. The users want the information and control provided at the user interface, but are not particularly interested in seeing the box that goes with it.

Hoffman prefers to use color in a minimalistic manner, thereby strengthening the colors' ability to draw attention. He believes that much of the product's aesthetic qualities can actually be achieved by using gray tones and that such an approach avoids a product design that is jarring in appearance. Hoffman welcomes more guidelines and standards on color use and believes that other designers inside and outside his company feel the same way. He cites as an example the need for standard colors for battery indicators, referring to a defibrillator that has indicators for battery charging, battery low, and charge done. He has found no established convention for the colors of such indicators and believes that continuing inconsistencies among manufacturers place an unnecessary burden on the user to remember product-specific color codes. One can imagine how the burden would increase for physicians and nurses who rotate among several hospitals.

A Physician's Viewpoint

Dr. Frank Block is a practicing anesthesiologist who has been affiliated with The University Hospital, Ohio State University in Columbus. He provides design consultation to patient monitor manufacturers and cochaired the Human Engineering Committee of the Association for the Advancement of Medical Instrumentation (AAMI). He describes the majority of color patient monitors in today's operating rooms as "garish."

> Color has been brought in by marketplace pressure. It is something that manufacturers use to get you into their booth at a medical convention, but it ends up being a nightmare to look at day after day.

Block accepts that color has value when it is applied intelligently, but laments the fact that

> many manufacturers leave it to the users to select the colors for the information on a display. You see displays that put red and blue together, colors that the eye cannot focus on at the same time. This causes fatigue.

Block is receptive to the development of standards for the color-coding of information on monitoring devices. However, reflecting on the state of design guidance in 1991, he commented,

> There really isn't a good reference source on the use of color in medical products.

Nevertheless, from the experience of using many different patient monitors, Block felt that a de facto standard was emerging for the color of blood pressure parameters, wave forms in particular: arterial pressure is displayed in red, central venous pressure is displayed in blue, and pulmonary artery pressure is displayed in green or yellow.

Block linked the choice of red for arterial pressure and blue for central venous pressure to the visual appearance of oxygen-rich blood and oxygen-poor blood, respectively. He linked the choice of either green or yellow for pulmonary artery pressure to the fact that early color displays provided limited color choices, which happened to include green and yellow. He believed it may be possible to develop and document several other de facto standards by which product manufacturers can make rational design decisions. Since 1991 Dr. Block has led the AAMI committee to develop rudimentary guidance on the use of color in medical device design (see Table 12.1).

Reflecting on the future role of color in patient-monitor displays, Block acknowledges that "there is a perceived benefit [of color] right now," but then poses the ultimate question: "Can you take better care of your patient by using a color display versus a monochrome display?" He would like to see research that would prove color's benefit to patient care, considering the added cost associated with color displays, but is not sure that the question could ever be answered satisfactorily.

A Success Story

In 1981 several manufacturers were producing their own colored labels for syringes. These labels were developed to prevent so-called syringe swapping, which occurs when a physician or nurse picks up the wrong syringe from a collection of syringes already prepared for drug injection. Unfortunately, despite good intentions, label colors were not consistent among manufacturers. Dr. Leslie Rendell-Baker, a professor of anesthesiology at Loma Linda University in California, recalls that

> there was no coherence across manufacturers in the color selection. One firm would use a green label for a given narcotic drug, one would use blue, and another would use red. This was confusing. The colored labels were not helping.

In response to the problem, Rendell-Baker led an ASTM subcommittee in efforts to address the problem, following up on

Table 12.1. *AAMI's suggested uses of color coding for medical devices.*

Color	Meaning
Red	• High priority level (warning) alarm • Emergency • Stop • Off • May be coded to physiological variables (e.g., arterial blood pressure) • May be coded to other standards such as anesthetic agent colors (e.g., red for halothane)
Yellow	• Medium priority level (caution) alarm • Low priority level (advisory) alarm (steady yellow) • Potential hazard • May be coded to physiological variables • May be coded to other standards such as compressed gas cylinders (e.g., yellow for air in the United States)
Green	• Start • On • Normal or ready • May be coded to physiological variables • May be coded to other standards such as compressed gas cylinders (e.g., green for oxygen in the United States)
Other Colors	• Any meaning except the above • May be coded to physiological variables (e.g., blue for central venous pressure) • May be coded to other standards such as anesthetic agent colors and compressed gas cylinders

Note—In some applications, specific standards may dictate other color coding requirements. For example, by convention, some devices use red to indicate ON and green to indicate OFF.

work started by the South African Society of Anesthetists. The group looked first at the legibility of drug names and dosages printed on labels for ampoules and vials. Their requirement for legibility was based on the "worst case" of a far-sighted person attempting to read the label at arm's length (20 in.) in the minimum permitted hospital corridor lighting (20 fc). They then looked at the options for color-coding user-applied drug-syringe labels. After considering the use of varying shades of a color— light, medium, and dark red, for example—to distinguish different members of a drug group and finding that this led to confusion, they established seven broad classes of drugs and assigned to each a unique color: yellow for anesthesia induction agents, red for muscle relaxants, blue for narcotics, salmon for tranquilizers, violet for vasopressors, and green for atropine and other drying agents. Their solution for color-coding antagonist drugs was to print diagonal white stripes alternating with the specific color for that group of drugs. Thus, narcotic antagonists have alternating diagonal blue and white stripes. The standard colors are defined in ASTM D-4774-88, *User Applied Syringe Labels for Anesthesia.*

Taking Initiative to Improve Color Use

One step that today's designers can take is to conduct a survey of their companies' products to analyze existing color conventions and design strategies. This simple and illuminating exercise will enable designers to document all the colors used in each product line (and in product development) and their associated meanings. Several inconsistencies will likely emerge from such a survey and may be traced to situations in which different products were designed by different designers and consultants (from both inside and outside the company) who gave insufficient regard to product-line integration. Also, inconsistent use of color may be found to be the result of aesthetics decisions that outweighed goals for consistency. The process of becoming more consistent in the use of color can start with the development of a corporate standard, or style guide, for color use. Such a standard may help guarantee that products from the same company are designed with consistent meanings assigned to colors. A corporate standard can be augmented by design guidelines on the effective use of color, such as how many colors to use and where to use them.

A broader challenge is to generate consistency among manufacturers regarding the use of color. To accomplish this goal, the development of guidelines, de facto standards, or formal industry standards may help. Although the idea of standards may annoy and frustrate the libertarians of the design world, the result is likely to be smoother interactions between medical products and their users. The initiative for a color standard may need to come from a consortium of medical product companies committed to improving the way colors used on their products communicate to users. Other associated, potential benefits may also result, such as reducing design time and cost and sharing liability for color selections.

Useful Information on Color

Designers may find the following resources useful.

- AAMI HE-1988, *Human Factors Engineering Guidelines and Preferred Practices for the Design of Medical Devices*. Association for the Advancement of Medical Instrumentation, 1988. This document presents human factors guidelines largely adapted from a military standard (MIL-STD-1472). Paragraphs 4.5 and 5.8 refer to Tables 3 and 7, which establish color codes for controls and transilluminated displays, respectively.

- ISO Draft Standard 9241-8, *Computer Display Color*. International Standards Organization, 1989. This draft standard includes a list of relevant literature, definitions for color terminology, guiding principles, performance requirements, design requirements and recommendations, and test methods. The standard examines carefully the physics of computer color displays and legibility issues. Tables 7.8.1 and 7.8.2 provide information about the luminance and hue contrast between colors, respectively, that aid in the selection of legible color combinations. In the draft standard Section 2, "Field of Application," states,

  > Color is most appropriate for images that are to be highlighted, located, differentiated, classified, associated, or segregated, or where a relationship is to be indicated, a quality or realistic appearance represented, a physical impression created or a concept coded.

- Pamphlet C-9, *Standard Color-Marking of Compressed Gas Cylinders Intended for Medical Use in the United States*. Compressed Gas Association, 1973. Defines the color codes established for compressed gas cylinders.

- *Using Computer Color Effectively* is a compendium on the use of color on computer displays. Chapter 11, "Computer

Color Guidelines," includes several useful figures and tables that can help designers choose effective colors, including:

—Table 11.1: Meanings of Colors for Display Applications

—Figure 11.5: Color Harmony Combinations

—Table 11.3: Discriminable and Legible Color Combinations on Achromatic Backgrounds

—Table 11.4: Easily Perceivable and Discriminable Colors for Thin Lines

—Table 11.5: Easily Perceivable and Discriminable Area-Fill

—Table 11.9: Comfort Ratings for Color CRT Text and Background Combinations

—Table 11.11: Ratings of Edge Sharpness of Color Backgrounds

- G. M. Murch, "Colour Graphics—Blessing or Ballyhoo," in *Readings in Human-Computer Interaction: A Multidisciplinary Approach.* Morgan Kaufmann Publishers, Inc., 1987, pp. 333–341. This article provides a useful set of guiding principles for effective color use.

- G. M. Murch, "Physiological Principles for the Effective Use of Color." *IEEE Computer Graphics and Applications,* November 1984, pp. 49–55. This article discusses color use in the context of compatibility with eye physiology.

- A. Marcus, "Designing Graphical User Interface." *UnixWorld* 7(10):137, 1990. This well-illustrated article provides examples of good and bad color combinations.

References

Albers, J. 1975. *Interaction of color.* New Haven: Yale University Press.

Marcus, A. 1986. Tutorial—The ten commandments of color. *Computer Graphics Today* 7(10):7–11.

Marcus, A. 1990. Designing graphical user interface. *UnixWorld* 7(10):137.

Marcus, A. 1992. Graphic design for electronic documents and user interfaces. ACM Press.

Smith, W. 1988. Standardizing colors for computer screens. In *Proceedings of the Human Factors Society 32nd Annual Meeting.* Santa Monica, CA: Human Factors Society.

Thorell, L. G., and W. J. Smith. 1990. *Using computer color effectively.* Englewood Cliffs, NJ: Hewlett-Packard and Prentice-Hall.

Wiklund, M. E. Unpublished results.

Chapter

13

The Role of Symbols in User Interfaces

The human brain is adept at recognizing symbols, which is why they are attractive for use on medical device labels and in software displays. Symbols are economic and efficient—one symbol can communicate a concept that might otherwise require several words or even a sentence to convey. This makes symbols especially useful in situations where information must be communicated but where space is limited—for example, on small devices and in computer-based displays.

Because they are not language dependent, symbols are especially appealing to companies developing products for international markets, such as the European Community. Using symbols in place of text can simplify product manufacturing and distribution; otherwise, companies must produce custom labels that use the language of the country importing the device. A design vision shared by many marketers is an entirely symbolic approach to hardware labeling, complemented by language-specific software displays that are relatively easy to modify. Symbols are not a panacea, however. The biggest problem with symbols is that they can be interpreted in a variety of ways. At the very least, this can create confusion, and if a symbol identifying a critical medical device function is misinterpreted, a patient injury or death could result.

Nevertheless, the potential for symbols to convey information quickly to a multilingual user population is compelling. Accordingly, the device industry is likely to move gradually from textual to symbolic user interfaces as more products are developed for international use. Therefore, manufacturers should place increased emphasis on producing professional-quality symbols by means of a systematic design process that incorporates user feedback. In addition, they should support the development of a robust, uniform vocabulary of medical device symbols.

Systematic Design

Generally, medical workers find symbols visually appealing but are ambivalent about their usability advantages—everyone has encountered a symbol he or she could not immediately decipher. For example, after glancing quickly at the sample symbol shown in Figure 13.1, people provide widely varying explanations of its meaning, ranging from a gas pump to an electric

Figure 13.1. *Example of an ambiguous symbol. Only when the symbol is placed in the context of an airline trip does its meaning—seat occupied—become clear.*

shaver. It is not until the symbol is placed in context (an airline trip) that people figure out that it means "seat occupied." The variety of responses underscores the fact that symbol interpretation is a subjective matter, and is why a systematic approach to symbol design is so important.

Early in the overall product development process, designers need to decide what role symbols will play in the user interface. Some manufacturers restrict the use of symbols to the communication of high-priority or patient safety-related information. Others use symbols more freely to communicate both important and routine information, or even to make the user interface more visually appealing.

In many cases symbol usage is dictated by international standards, current industry practice, or established corporate style guides, which can make the process of selecting symbols an easier but perhaps more frustrating task. Designers should document decisions on symbol use to ensure consistency and prevent later corruption of the user interface because of arbitrary decisions to add or remove symbols for cosmetic reasons.

Decisions about symbol use can be validated by a design review involving three to five representative users who understand how a product will be used and how symbols are used on similar equipment. At this stage designers should also investigate how specific symbols might be produced (e.g., displayed on a screen, silk-screened or adhered to a panel, or engraved) so that constraints regarding visual appearance, size, resolution, readability, and durability are understood.

Graphic and Pictographic Symbols

When designers need new symbols, they can take one of two general approaches: graphic or pictographic (Cushman and Rosenberg 1991, 131–132). A *graphic symbol* is an abstract or arbitrary symbol used to represent a concept that may not have a real-world analog, such as an exclamation point inside a triangle used to mean "see instruction leaflet" (Figure 13.2a). In fact, many people would assume that this symbol indicates a hazard, considering that its standard meaning in other domains is "safety alert." Such symbols do not convey useful information until people learn exactly what they mean. This makes graphic symbols less desirable for medical device applications that require users seeing a symbol for the first time to understand what it means immediately.

By comparison, a *pictographic symbol* is an analog representation of a familiar object or action. The umbrella symbol (Figure 13.2b), which indicates that an object should be kept dry, is a good example of a pictographic symbol, as is the broken glass symbol used to convey the notion that an object is fragile (Figure 13.2c).

If pictographic symbols are well designed, their meaning is intuitively obvious. However, as Cushman and Rosenberg point out,

> The designer [of the symbol] assumes that the user has had some relevant experiences with the objects. If the validity of this assumption is questionable, verbal labels should also be provided. This redundancy facilitates learning and makes the product easier to use, especially for infrequent users. . . . Pictographic symbols should be used when the intended product users do not share a common language and when use of multilingual labels or different labels for each language is impractical (Cushman and Rosenberg 1991, 131–132).

Once designers have chosen the appropriate approach, they should sketch a variety of symbols with no particular regard for aesthetics. Instead, they should focus on discovering and combining visual elements (i.e., representations of objects and actions) that are likely to convey an intended message. During their sketching exercise, which can be performed by hand or on computer, designers should keep in mind the

Figure 13.2. *Graphic symbol meaning (a) "see instruction leaflet"; pictographic symbols meaning (b) "keep dry" and (c) "fragile."*

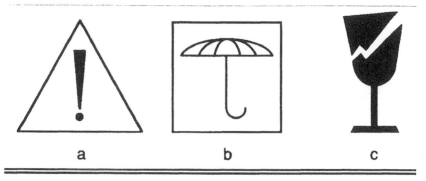

a   b   c

available graphic possibilities. For example, symbols displayed on a computer monitor may allow designers to use color to communicate information—such as red for hazard—and may incorporate dynamic features, such as a flashing element indicating a change in the condition of an object.

The initial sketching exercise may produce 10 or more sketches for each information element, ranging in quality from unacceptable to very promising. The sketches can be evaluated informally by team members to determine, say, the three most promising designs, which can then be refined until they are suitable for use on the actual product. At this point designers should concentrate on establishing rules to assure consistent size, proportion, line quality, fill pattern, and color so that the symbols have a common look—as if they are part of the same family.

Regarding the design of icons, graphic designer Aaron Marcus of Aaron Marcus + Associates (Emeryville, CA) advises,

[The use of] any extraneous decorative parts should be carefully weighed against the confusion they may cause the viewer. On the other hand, the icons should not be so simple that they all seem identical; they should be distinguishable.

The visual differences among icons should be significant, Marcus says, because random or idiosyncratic changes merely take more time to figure out and create the possibility of misinterpretation (Marcus 1992, 63). Additional design guidelines, adapted from several sources*, include the following:

- Avoid extraneous detail.
- Avoid too many parts.
- Design first in monochrome, then add color.
- Limit the number of colors.
- Limit size variations.
- Limit variations in line width.
- Use a constant scale to represent similar size objects.
- Use the same visual elements consistently to represent the same object or action.
- Establish a consistent figure-to-ground relationship.
- Limit variations in the ratio of white space to ink.

---

*Adapted from the following sources: Cushman and Rosenberg 1991; Marcus 1992; and Blattner, Sumikawa, and Greenberg 1989.

Getting Users Involved

At this point in the symbol design process, potential users can help identify problems with symbols and determine which designs communicate best. One approach is to show individual users a symbol for a few seconds and get their first impression of what it means. If you have developed alternative designs for each symbol, obtaining an unbiased impression may require a large number of potential users. After they provide individual first impressions, groups of users can compare and discuss design alternatives aloud, identifying those they prefer most. Users also can rate each symbol—for example, on a scale of 1 to 5—or rank them from least preferred to most preferred. This approach makes it easier for designers to quantify user preferences.

Before making a final selection, designers should consider whether a set of symbols, such as eight symbols on a control panel, will function well as an integrated set. For example, two-symbol designs may each be effective when viewed separately, but may create ambiguity when viewed together. In addition, the number of symbols displayed together affects how conspicuous each symbol is and whether it communicates effectively. For instance, a control panel containing a dozen symbols would likely frustrate users and make them feel as if they were taking an IQ test, especially if some symbols are unfamiliar. Conversely, users might think a control panel with only a few symbols and a large number of text labels is old-fashioned and inappropriate for quick, intuitive use.

For purposes of design quality and liability protection, the design team can validate final symbol designs using an acceptance criterion developed by the American National Standards Institute (ANSI) for the evaluation of safety symbols. According to ANSI guidance, users should be able to correctly define the meaning of critical symbols 85 percent of the time, with no more than 5 percent instances of critical confusion (ANSI Z535.3-1991). In a less-demanding alternative approach that is, nevertheless, well suited to the evaluation of noncritical symbols, potential users judge the appropriateness of symbols to their intended uses by matching symbols with possible meanings (Nolan 1989, 380–384). A high proportion of mismatches indicates the need for redesign.

Current Symbol Vocabulary

It is not clear how well the current set of medical device symbols would fare if subjected to user testing. Seth Banks, manager of market communications and industrial design with GE Medical Systems (Waukesha, WI), says,

> I'm not aware that there has been any substantial user pretesting of the symbols already in common usage.

Banks feels that the medical device industry is moving inexorably toward greater use of symbols, primarily as a response to manufacturer expansion into international markets. As a contributor to Working Group 5 of International Electrotechnical Commission (IEC) Subcommittee 62A, which deals with symbols and ergonomics for electrical equipment used in medical practice, Banks is concerned that symbol vocabulary could proliferate until it is out of control. As an example, he cites the recent work item proposal (SC 3C/SC 62A, October 1992) by the German National Committee of IEC to adopt for standard use approximately 85 new symbols for nuclear medicine, radiological, ultrasound diagnostic, and general medical devices. Banks says,

> These symbols attempt to create an entirely new visual language. We in the United States are basically symbol illiterate—our culture has been brought up using words, not symbols, to communicate. A sudden, dramatic shift to symbolic labeling on medical devices could be problematic. Dramatically increasing a symbolic vocabulary will make things more difficult for users and will reduce their speed in recognizing symbols. I would like to see more standard symbols than we have today, but we should move ahead slowly.

Meanwhile, Banks's design staff has taken the initiative to use more symbols on their user interfaces—once the designs have passed rigorous testing by GE.

Peter Carstensen, associate director at CDRH's Division of Small Manufacturers Assistance, shares many of Banks's concerns. As convener of Working Group 5, he expects the group to "look at the big picture" regarding the use of symbols on medical devices.

I understand the desire among manufacturers, especially those in Europe responding to the EC '92 initiative, to avoid using different control panels on their products. There has to be a lot of economic pressure to produce a single, multilingual product. But I don't think the set of symbols proposed by the Germans is workable—because, among other things, far too many of the symbols are not intuitive. I hope the working group can create a more usable set. Specifically, I would like the national delegates to gather input from the marketplace regarding the proposed symbols. Then, I would like to work with individuals on the committee, and their design resources, to refine the symbols and conduct user testing.

Greg Welyczko, manager of medical and industrial standards at Ohmeda (Madison, WI), a manufacturer of anesthesia systems, also participates in symbol design as secretary of Working Group 5. He feels that enlarging the medical device symbol vocabulary beyond a limited number of safety- and hazard-related symbols is problematic.

The symbols proposed to IEC may be subject to misuse and misinterpretation. Any symbol used on a medical device must be extremely intuitive and easy to interpret correctly. Otherwise, you must use words in addition to the symbols. Therefore, why bother with symbols?

Regarding any expansion of the symbol vocabulary, Welyczko says,

Not only are the symbols I have seen confusing, they are specific to a certain kind of medical device. Symbols suitable for a radiology device will not be suitable for an anesthesia device. Yet, symbols are often presented as if they are for general use on any kind of medical device. This opens the door for misapplication, which will frustrate users. The standards had better state clearly what type of device a symbol applies to.

Conclusion

Will future medical device interfaces use an increasing number of symbols? The mounting international pressure suggests so. However, the last thing medical workers need is to have to play guessing games regarding what user-interface symbols mean.

Given the pressure and stress medical workers encounter, this would surely lead to decreased productivity and user errors, and could have serious implications for patient safety and the financial well-being of medical institutions. That is why designing effective symbols is so important, particularly if a symbol serves a critical function (see Table 13.1).

Users should be able to look at a symbol and understand immediately and unambiguously what it means. Unfortunately, symbols are inherently ambiguous. This is why following a systematic design process, enriched with abundant user feedback, is more important than ever before. Such a process ensures that manufacturers will be able to create symbols that reliably communicate the same basic message to the vast majority of users, leaving little room for serious misinterpretation.

Table 13.1. *Publications describing the use of symbols applicable to medical devices.*

General Principles for the Creation of Graphic Symbols, ISO 3461-1. Geneva: International Organization for Standardization, 1988.

General Principles for the Creation of Graphic Symbols for Use on Equipment, IEC 416. Geneva: International Organization for Standardization, 1988.

Graphic Symbols—Use of Arrows, ISO 4196. Geneva: International Organization for Standardization, 1984.

Graphical Symbols for Electrical Equipment in Medical Practice, IEC 878. Geneva: International Organizational Electromechanical Commission, 1988.

Graphical Symbols for Use on Equipment, IEC 417. Geneva: International Organization for Standardization, 1987.

Graphical Symbols for Use on Equipment—Index and Synopsis, ISO 7000. Geneva: International Organization for Standardization, 1989.

Safety Colors and Safety Signs, ISO 3064. Geneva: International Organization for Standardization, 1977.

These publications can be ordered from the American National Standards Institute (ANSI), 11 W. 42nd St., 13th Floor, New York, NY 10036; 212-642-4900.

References

Blattner, M., D. Sumikawa, and R. Greenberg. 1989. Earcons and icons: Their structure and common design principles. *Human-Computer Interaction* 4(1):21–22.

Criteria for safety symbols. 1991. ANSI Z535.3-1991. Washington, DC: National Electrical Manufacturers Association.

Cushman, W., and D. Rosenberg. 1991. *Human factors in product design.* New York: Elsevier Science Publishers.

Marcus, A. 1992. *Graphic design for electronic documents and user interfaces.* New York: ACM Press.

Nolan, P. 1989. Designing screen icons: Ranking and matching studies. In *Proceedings of the Human Factors Society 33rd Annual Meeting.* Santa Monica, CA: Human Factors Society.

Chapter

14

Communicating Clinical Information with Auditory Signals

As you walk down a hospital hallway, you hear three beeps coming from behind a closed door. It could mean any number of things: a digital thermometer has just completed measuring a patient's temperature, a patient-monitor alarm has been tripped, a doctor is being paged, or the popcorn in the lunch-room's microwave oven is done. Determining the real meaning of a sound requires locating its source or paying close attention to other cues, such as its tone, rate, or volume. Usually, an experienced listener has little trouble sorting out the actual meaning, but the sorting process may increase a person's cognitive work-load at a time when it is already high. In part, the problem may be that many of today's medical devices produce a limited range of sounds—usually beeping noises—that are relatively ambiguous compared to more-distinct auditory signals, such as a sequence of tones or the spoken word.

Some product developers and users advocate increasing the auditory vocabulary of medical devices in order to facilitate information flow. For example, one researcher showed that playing a few bars from the song "I Left My Heart in San Francisco" in place of a beep was an effective method of signaling cardiovascular distress in a patient (Block 1991, A497). Aviation research has also established the usefulness of more-complex auditory signal coding in aircraft cockpits (Patterson and Milroy). The tendency of product developers to embrace new technologies such as audio chips (microprocessors that generate sound) may satisfy those who seek to expand medical device vocabularies. Other designers and users, however, would settle for greater consistency in the use of a limited set of auditory signals and would prefer to avoid more-complex vocabularies, which they feel could become a distraction.

Appropriate Applications

There are several important guidelines for effectively using auditory signals (versus, say, visual displays) to communicate information. Consider using sound when

- The origin of the signal is itself a sound (e.g., a heartbeat).
- The message is simple or short.
- The message will not be referred to later.
- The message deals with a discrete event in time (e.g., an arrhythmia).
- The message is a warning that calls for immediate action.
- The message presents continuously changing information.
- Visual or speech systems are overloaded.
- Illumination limits the use of vision.
- The person receiving information moves from one place to another (McCormick 1976, 123–124).

These guidelines stress the importance of using auditory signals in applications that do not place high demands on the user's memory. This constraint applies especially to auditory signals that do not recur. Unquestionably, medical workers have little patience for a medical device that gives them but a single opportunity to acquire important information and taxes their powers of recall, thereby increasing the chance of mental error.

The guidelines are striking in the degree to which they conform to actual conditions of medical device use (e.g., nurses moving about the intensive care unit or an anesthesiologist moving between the anesthesia workstation and the patient). Perhaps recent advances in auditory signaling technology will increase the number of appropriate applications in the medical device inventory.

An Innovative Application

The Nellcor N-100 pulse oximeter (Figure 14.1), introduced in 1983 by Nellcor, Inc. (Pleasanton, CA), exemplifies how a relatively simple auditory signal can clarify a complex situation. The product is used to determine if a patient is getting enough oxygen and also provides information about heart function. It provides a digital readout of a patient's oxygen-saturation level, measured in terms of the percentage of hemoglobin-carrying oxygen and the pulse rate. A noninvasive device that clips on the patient's finger is used to make the measurements. The digital readouts are augmented by a variable-pitch beep synchronized to the patient's heartbeat. When a patient's oxygen-saturation level drops, so does the pitch of the beep, alerting the

Figure 14.1. *Nellcor N-100 Pulse Oximeter.*

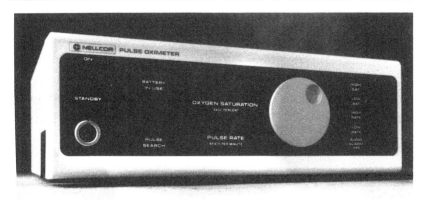

Photo courtesy of Nellcor Incorporated. Pleasanton, California.

user (typically the person administering anesthesia) to a potentially life-threatening condition. If there is an arrhythmia, the clinician hears a change in the rhythm of the beeps in real time.

The innovative use of sound to convey clinical information arose when Nellcor observed that a practitioner cannot always look at the pulse oximeter's digital displays, either because he or she is focusing on the patient or because the device is out of view. The variable-pitch, tone beep enables the practitioner to detect changes in oxygen saturation or heart rate, as well as arrhythmia, without looking away from the patient. According to David Swedlow, M.D., vice president of medical affairs and technology development at Nellcor, the variable-pitch beep has been well received by the anesthesia community and has been a major factor in the success of the company's product.

> The ability to detect changes in the patient's condition is particularly important during intubation, when one is not always certain about the depth of anesthesia and arrhythmias are common.

Nellcor holds a patent on the variable-pitch feature, but it has licensed the technology to several other patient-monitor manufacturers.

Remembering the N-100's introduction, Swedlow says,

> Users would listen to the variable-pitch beep and ask: "Why is it doing that?" Even after they received an explanation, some users still considered the feature silly. But many others really liked it, and it caught on.

Even a feature that many consider useful can create problems, however.

> A possible downside is that surgeons can become so attuned to the sound of the pulse oximeter that a change in pitch can divert their attention from the surgical field.

Swedlow believes the effectiveness of the variable tone depends on limiting the use of sound in the clinical environment. For example, he is against using sound for less important clinical purposes (e.g., electronically mimicking the patient's breathing pattern with a "swish" noise). He also sees a problem using sound to indicate a condition for which there is no analogue, such as an excessive concentration of anesthetic vapor.

We do not have physiological equivalents to the sound of a muffler falling off a car—a distinct sound that drivers can easily recognize.

Achieving Consistency

The effective use of sound in medical devices requires more than a clever application, however. It is important to use sound consistently; otherwise, a sound can produce confusion and negative transfer (when a user incorrectly assigns a meaning to a sound based on past experience with a similar device). Ideally, different brands of the same type of device would emit approximately the same sound. In pursuit of this ideal, the American Society for Testing and Materials (ASTM) formed Subcommittee F29.03.04 on harmonization of alarms to develop a specification for alarm signals, including audio signals, for medical equipment used in anesthesia and respiratory care. The committee's progress includes development of a draft specification that was distributed for the purpose of committee review only.

It seems, however, that the committee has hurdles to overcome. Swedlow has observed its activities with great interest and a measure of pessimism.

> While manufacturers are uniformly desperate to achieve harmonization, users are sharply divided on the issue of how best to use sound. One group is committed to a complex matrix of sounds to indicate a large range of alarm conditions. The other group advocates limiting the use of sound to indicate only high-, medium-, and low-priority events. I expect that the committee will not be able to resolve their differences and that we will not see a specification.

Swedlow agrees with the latter group and would like to see manufacturers adopt an approach similar to the one implemented by North American Dräger, Inc. (Telford, PA) on its anesthesia workstations.

The Narkomed 4 Anesthesia System, for example, emits distinct auditory signals (called warnings, cautions, and advisories) for high-, medium-, and low-priority events (see Table 14.1). To identify the specific nature of the alarm, users turn their attention to drop-down windows on a computer display located on the workstation or on a pivoting boom.

Table 14.1. *Auditory signals for Narkomed 4 Anesthesia System.*

Category	Auditory Signal	Sample Alarm
Warning	Sequence of tones followed by a continuous tone[a]	Apnea for 30 seconds
Caution	Intermittent tones[b]	Pulse > high-alarm limit
Advisory	Single tone	O_2 sensor cord disconnected

[a]After an initial burst (tone) at full volume (when the alarm first occurs), there is a 6-second pause. This is followed by a burst at one-third volume, a 5-second pause, a burst at two-thirds volume, a 4-second pause, then a burst at full volume. After that, there is a 3-second pause followed by a full-volume constant burst until the alarm condition is removed.

[b]A three-pulse burst is repeated every 30 seconds until the alarm condition is cleared.

Source: North American Dräger, Inc.

Designing Effective Auditory Signals

Designing effective auditory signals calls for a structured approach similar to that used to design icons, the graphic elements found in modern software applications. According to Blattner et al.,

> Even though they utilize different senses, icons and earcons [auditory icons] involve similar communication needs and design problems. Knowledge obtained by examining theories of icon design can be used to model equally successful approaches for [auditory icons] (Blattner et al. 1989, 21–22).

As is the case with visual icons, auditory icons vary from representational to abstract. For example, the caricatured sound of a person snoring would be a representational icon for sleeping, while a low-pitch buzzing sound would be more abstract. First-time listeners are bound to derive greater meaning from representational auditory icons than from abstract ones. However, an abstract sound is likely to be perceptibly simpler and less irritating to listen to on a repetitive or ongoing basis.

According to Blattner et al., informative sounds can also be categorized as symbolic, nomic, and metaphorical.

Symbolic mappings rely on social convention for meaning, such as applause for approval; nomic (the most concrete) representations are physical, such as the sound of a closing metal cabinet for closing a file; and metaphorical mappings are similarities, such as falling pitch for falling object (Blattner et al. 1989, 22).

Combining taxonomies, the variable tone emitted by the Nellcor N-100 is an abstract as well as a metaphorical auditory icon.

Blattner et al. also refers to one-element and compound auditory icons; these terms refer to words and phrases, respectively. A one-element auditory icon might be a digitized sound, a beep, or a short succession of tones (akin to the letters in a word) that convey a specific meaning, such as *alert*. A compound auditory icon (created from several one-element auditory icons) might produce the phrase *alert blood pressure high*. The key to developing useful compound auditory icons is to start with a primary set of one-element auditory icons that are easy to distinguish both from one another and from ambient noise.

As is the case with other aspects of the user interface, developers of auditory icons should expose users to preliminary auditory signals to get their reactions and suggestions for improvements. Such evaluations could be modeled after approaches to icon or warning-symbol evaluations, which record users' interpretations of a graphic after a short, initial exposure to it (see chapter 27).

Using Synthesized Speech

Voice output is becoming more popular in applications that call for precise and reliable communication when a visual display is insufficient. For example, one manufacturer has equipped a voltmeter with a voice synthesizer that will verbally read out voltage levels at the press of a button, a handy feature for technicians who wish to keep their eyes on their work while working with live electrical wires.

One can imagine several potentially beneficial uses of voice output in the medical area. Perhaps in the future a defibrillator equipped with a voice synthesizer could pronounce the sound ready when fully recharged so that the clinician can focus on the proper paddle placement on the patient's torso. Perhaps a patient monitor could announce certain alarm conditions, such as apnea, or read out certain parameter values on request, again so that the clinician can direct his or her full attention to the patient instead of the monitor.

Will synthesized speech one day replace compound auditory icons? Probably not, because there are drawbacks to synthesized speech. For instance, voice output has a tendency to annoy people. Drivers hated cars of the early 1980s that verbally commanded them to put on their seat belts and indicated when a door was ajar. Also, a warning in the form of synthesized speech might be difficult to distinguish in an environment filled with human conversation (e.g., an emergency room). Another problem with using voice is selecting a language suited to all users. Manufacturers already have enough difficulties coping with the need to produce user manuals, display text, and control labels in several foreign languages.

Conclusion

The ultimate guideline for using auditory signals in the user interfaces of medical devices is to use them appropriately and sparingly. Clinicians dislike medical devices that take their focus away from the patient. After all, some clinicians will even get annoyed with other people talking, let alone a medical device talking in a slightly robotic tone.

Nevertheless, the Nellcor N-100 demonstrates how the limited use of an auditory signal has produced major benefits. Certainly, there are many more medical applications to explore, particularly as a result of the advent of digitized sound and voice synthesis. Also, there are many existing applications that can be improved in terms of information content and harmonization with other auditory signals. In all cases users should be an integral part of the development process and should judge the usefulness of auditory signaling approaches in carefully conducted user tests.

References

Blattner, M., D. Sumikawa, and R. Greenberg. 1989. Earcons and icons: Their structure and common design principles. *Human Computer Interaction* 4(1):21–22.

Block, F. 1991. Evaluation of physicians' abilities to recognize musical alarm tones. *Anesthesiology* 75(3A):A497.

McCormick, E. 1976. *Human factors in engineering and design.* New York: McGraw-Hill.

Patterson, R. D., and R. Milroy. *Auditory warnings in civil aircraft.* MRC Applied Psychology Unit, Civil Aviation Authority, Contract Report Number 7D/S/0142.3.

Chapter

15

Designing Medical Devices for Older Users

America is growing older. A special issue of *Human Factors* devoted to the topic of aging states:

> In the decade between 1990 and 2000, the number of persons aged 55 and over will increase by 11.5%, a gain of nearly 5 million people (U.S. Bureau of the Census, 1988) (Czaja 1990, 505).

According to forecasts, by the year 2020, one out of every five or six Americans will be older than age 65.

As they grow older, many people experience a decline in their physical and mental abilities, with an accompanying reduction in their capacity to perform product-related tasks. Enabling these people to continue to use technology effectively, particularly medical devices intended to improve or preserve their health, is an important challenge to manufacturers.

As the ranks of the elderly increase, greater demands will be placed on the nation's health-care system to enable people to live independently longer. New health-care strategies involving more self-care in the home are being developed. Already, many people of varying ages routinely use medical devices to measure their temperature, pulse, blood pressure, and blood-glucose level. It seems likely, therefore, that as additional technologies are refined and become more affordable, more medical devices will be used in the home by older individuals. The efficacy and safety of these devices will depend largely on their usability, particularly for older individuals coping with both physical and mental limitations. Accordingly, it serves the interests of both end users and manufacturers of medical devices for the latter to increase the attention they pay to the needs of older consumers.

Understanding the Marketplace

Robin Barr, chief of the cognitive functioning and aging section of the National Institute on Aging (Bethesda, MD), cautions designers against a common form of myopia: designing products to meet their own needs as opposed to those of the market.

> If you are designing for older users, you cannot assume they will have the same sensory and cognitive capabilities you do. Therefore, designers should open their minds to the needs of real users and conduct appropriate design studies to ensure that those individuals' needs are being met.

Joseph Koncelik, professor of industrial design at The Ohio State University (Columbus, OH) and a specialist on designing products for older people, agrees, adding:

> I constantly confront designers in my classroom who use themselves as a model for others.

Koncelik thinks that the stereotype of the technically incompetent older consumer will change rapidly over the next few years. He characterizes the elderly population as

> an increasingly educated group, many of whom have college degrees and who are completing successful careers.

He projects that older consumers

> will be increasingly active, healthy people who demand products that meet their needs. They will be a strong rather than a

docile market force, and any sales organization that operates on the stereotype that older people are infirm and dependent on others is in for a big surprise. Older consumers are going to be very selective about the products they purchase.

Koncelik says that older consumers will accept a new technology if it provides a real benefit, but will be less likely to accept it if it offers no particular benefit over existing technologies.

Cognitive Limitations

Regarding the special needs of older persons, Barr says,

> Older people have different health problems from those of younger people, and designers need to take these into account. For example, the cognitive abilities of older people vary considerably.

While many remain sharp, others experience attention deficits and what has been called "cognitive rigidity," a condition that makes it difficult to learn new procedures. According to Barr,

> Devices such as a VCR or a medical device of equal complexity are a problem because they require a large number of procedural steps.

He advises manufacturers to reduce the number of steps in a given procedure whenever possible, echoing the findings of an FDA study on human errors in blood-glucose-measurement devices (see chapter 5).

Koncelik recommends building redundant cues into a design as a way to facilitate user tasks for older people; for example, he suggests designing controls that provide visual, audible, and tactile feedback. He proposes designing display sequences that include prompts, such as asking users if they are sure they want to initiate a given action. Such prompts, which are commonly found in application software for personal computers, give users a chance to change their minds or fix an error.

Regarding the marketing of medical devices, Koncelik suggests providing personalized training with products when appropriate.

> Manufacturers are mistaken if they feel they can deliver a more-complex product with a set of instructions and expect a majority of people to use it successfully, even if the instructions

are well designed. They should seek opportunities to show people how to use the product, such as demonstrating it at outpatient clinics and senior centers.

As an alternative to in-person demonstrations, Koncelik suggests that manufacturers include a demonstration videotape with the product.

> The strength of a videotape is that people can watch it repeatedly at their own pace. A videotape's obvious weakness is that people cannot ask questions, although a manufacturer could set up a toll-free number so that people can call in their questions. It turns out that older people are actually better long-term learners than younger people, even though it takes them longer to learn.

Barr attributes some of the difficulties older people have coping with a technology to a lack of familiarity with its features. In fact, studies have shown that one of the best ways to improve learning and task performance by older individuals is to design a product with familiar features.

> Today, people are familiar with automatic washing machines and know how to use them. When they first came out, some people undoubtedly had difficulty figuring out how they worked. I fully expect today's computer hackers to continue using computers effectively into their old age. However, the same people may have trouble using tomorrow's technology if it requires methods of interaction that are unfamiliar to them.

Familiar product features and modes of interaction for medical devices used by older people include simple readouts, tactile buttons, rotary knobs, and toggle switches, rather than complex displays, trackballs, and scrolling lists, which may be better suited to the next generation of older users.

This is not to say that older individuals cannot cope with unfamiliar technologies. In fact, one study found that older individuals actually derive considerable value and mental challenge from interacting with computer technology, concluding that

> older adults are willing and able to use computers in their own homes if the system is simple, features are added in an incremental fashion, and they are provided with a supportive environment [i.e., one-on-one training] (Czaja et al. 1990, 146–148).

In addition to providing familiar features and modes of interaction for older users, designers can maximize learning and task performance by ensuring compatibility between the stimulus (cues or directions) provided by a device (and its user manual) and the response required of the user. Consider, for instance, the hypothetical case of a medical device with instructions telling users how to place their index finger over a sensor to obtain a physiological measurement. In place of a text-only instruction, a manufacturer might provide a diagram showing a user's hand with the index finger extended and a directional arrow leading from the finger tip to the sensor, establishing a strong compatibility between the instruction (stimulus) and user action (response).

Sensory Limitations

Impaired vision and hearing are examples of sensory limitations common among older persons. Such impairments can be measured using eye charts and audiograms and can be resolved in part by corrective lenses and hearing aids. Nevertheless, their existence requires that specific design features be incorporated into medical devices if they are to be used effectively by older persons.

Figure 15.1 shows the decline in the general population's visual acuity as a function of age. The percentage of individuals with a static visual acuity of 20/50 or worse, for instance, increases from 9 percent for individuals 55 to 64 years of age to 30 percent for individuals 75 to 79 years of age (Small 1987, 499–500). The obvious implication for designers is to use somewhat oversized fonts for the displays, readouts, and labels on devices designed for older users.

Older persons may also lose their ability to see well in low light, so products used in reduced-lighting conditions should include some form of supplemental illumination if it is necessary to read a display or manipulate controls.

Some older persons experience a loss in their ability to differentiate light intensities and colors. For example, they may have trouble discriminating the edges on green, blue, or violet objects (COMSIS Corp. 1988, 22). Therefore, although these colors may be acceptable choices for background features, they should not be used in the foreground—in text or graphic objects, for example.

Figure 15.1. *Percentage of population with a static visual acuity of 20/50 or worse.*

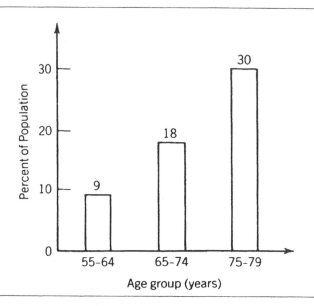

From Salvendy (1987), *Handbook of Human Factors*. Reprinted by permission of John Wiley & Sons., Inc.

Koncelik advises taking care to minimize reflected (specular) glare off device surfaces because the increased opacity of the eye's lens (the natural change that ultimately leads to cataracts) makes older individuals particularly sensitive to glare. This finding suggests that special design measures should be implemented, such as matte finishes on control panels and antiglare coatings on displays. Koncelik also suggests using inversed displays, which present white characters on a black background, thereby reducing the amount of light entering the eye.

Figure 15.2 shows the decline in the general population's hearing ability as a function of age. The chart, which combines hearing-loss data for men and women, shows how hearing tends to decrease as sound frequency increases. Data on hearing loss for men versus women show that, as they age, men suffer greater hearing loss at frequencies in the 3000–6000-Hz range, while women suffer greater losses in the 550–1000-Hz

Figure 15.2. *Increases in hearing impairment as a function of age.*

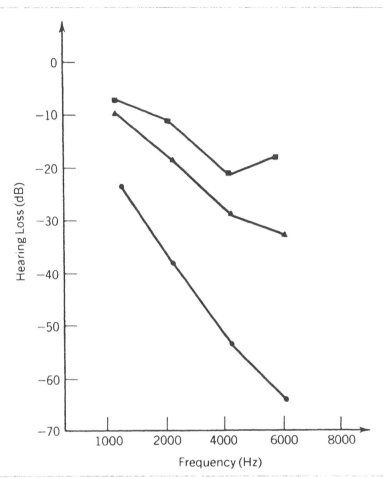

From Salvendy (1987), *Handbook of Human Factors.* Reprinted by permission of John Wiley & Sons., Inc.

range. This suggests that older persons will be most sensitive to audio signals in the moderate range (about 1500 to 2500 Hz). A study of the abilities of older persons to detect the sound of a ringing telephone in varying acoustical environments concluded that

> the overall superiority of [an] electronic bell ringer [was] probably due to the peak sound energy from 1000 Hz to 1600 Hz, a

sensitive range for both the young and old subjects as verified by their audiograms (Berkowitz and Casali 1990, 132–136).

This finding may be significant to manufacturers developing medical devices incorporating audio signals, such as the beep emitted by an electronic device to indicate that a measurement has been completed.

Slowed reaction time is a sensory limitation that is perhaps not as easy to accommodate in device design as is impaired sight or hearing. Reaction time continually increases (i.e., degrades) from an optimum achieved in early adulthood. Studies show an increase in reaction time of about 20 percent by age 60 for simple reactions, such as detecting a visual signal, and larger increases for more-complex signals, such as detecting a specific auditory signal embedded in a sequence of other signals (Small 1987, 499–500). This suggests that it is important for medical device manufacturers to develop user interfaces that do not require older users to respond rapidly to stimuli or to perform tasks with precise timing.

Physical Limitations

Designers of medical devices for older users should be aware that a significant number of people lose 10 percent to 20 percent of their overall strength by the time they reach 60 to 70 years of age, with a greater percentage of loss thereafter. Also, their mobility may be limited by joint disease and its associated pain.

These limitations suggest that medical devices for the elderly incorporate controls with large-diameter knobs so that rotation requires less fine motor control, textured knob surfaces that requires less pinching strength to eliminate finger slippage, and a low resistance to motion so that turning requires less exertion. Alternatively, a manufacturer might use a control that does not require turning, thereby eliminating a potentially painful motion. Anthropometric data on older persons, a necessity for developing physically accommodating and so-called transgenerational or universal designs, are readily available in the human factors literature.

Conclusion

The process of designing medical devices for use by older individuals is not particularly different from the process of designing for other special populations, such as children, young adults, or individuals with disabilities. The same fundamental principles of understanding the user's needs and validating designs through user testing apply.

Designers who take the initiative to address the needs of older consumers will find their efforts supported by a growing body of knowledge about these people. Moreover, the design features that especially benefit older persons will find favor with other users as well. When knobs are easy to turn, glare on displays is reduced, and instructions are presented in a clear and simple manner, everyone benefits.

Recommended Reading

Koncelik, J. 1982. *Aging and the product environment.* Florence, KY: Scientific and Academic Additions.

References

Berkowitz, J., and S. Casali. 1990. Influence of age on the ability to hear telephone ringers of different spectral content. In *Proceedings of the Human Factors Society 34th Annual Meeting.* Santa Monica, CA: Human Factors Society.

COMSIS Corp. 1988. *Product safety and the older consumer—What manufacturers/designers need to consider,* CPSC Publication 702. Washington, DC: Consumer Product Safety Commission.

Czaja, S. 1990. Special issue preface. *Human Factors* 32(5):505.

Czaja, S., M. Clark, R. Weber, et al. 1990. Computer communication among older adults. In *Proceedings of the Human Factors Society 34th Annual Meeting.* Santa Monica, CA: Human Factors Society.

Small, A. 1987. Design for older people. In *Handbook of Human Factors,* edited by G. Salvendy. New York: John Wiley.

Chapter

16

Cumulative Trauma Disorders: Implications for Product Design

Good engineering practice (not to mention the fear of liability claims) drives product developers to conduct extensive health and safety analyses of their product designs. Typically, such analyses focus on possible physical threats to the user's body— such as electricity, energy beams, chemicals, and moving parts—and may lead to important design changes and the inclusion of warning labels. But rarely does a product warn users to wear protective splints on their wrists or have a label that states: Repetitive use of this device may cause permanent injury (Figure 16.1). This is because the health and safety implications of using products that require repetitive body movements are only now coming to light. It turns out that some seemingly harmless products (e.g., a keyboard or a pair of pliers) are capable of causing cumulative trauma disorders (CTDs)

Figure 16.1. *Hypothetical warning concerning the hazard of repetitive hand motions.*

⚠️ **WARNING**

Repetitive use of this device may permanently injure your wrist.

that may require corrective surgery. In addition to repetitive motion, CTD has been linked to repeatedly overexerting muscles, applying too much force to a part of the body, assuming awkward postures while performing tasks, and not resting enough between periods of high activity. Patient treatments include extended rest, medication, splints, and surgery. However, CTD victims may still suffer lifelong effects caused by poor product design and work habits.

The insidious nature of CTD injuries, which may take months or even years to develop, makes them easy to overlook during a traditional health and safety analysis. However, CTD is a health risk that warrants close attention by product designers, particularly in cases where a product requires extensive manipulation. Much can be done to design a handheld or hand-operated device to minimize users' exposure to cumulative traumas. For example, the axis of a handle can be realigned to reduce the frequency and extent of wrist deviations from a neutral position; padding can be added to a handle to better distribute the mechanical force applied to the hand; spring action or levers can be incorporated in a design to reduce the necessary gripping forces; the work-surface height can be adjusted to afford better body posture. Such design changes, intended primarily to protect users' health, may also improve product usability, leading to greater user comfort, satisfaction, and productivity. Accordingly, product manufacturers should include in their health and safety studies an evaluation of CTD risks and they should implement design changes where necessary to protect users.

Magnitude of the Problem

Cumulative trauma disorder is most prevalent among people whose occupation requires them to perform the same task repeatedly. *Business Week* reports,

> [Repetitive stress injuries] in offices and factories are responsible for 56% of all workplace illnesses—185,000 reported cases in 1990. Workers' compensation claims and other expenses from these injuries may cost employers as much as $20 billion a year, estimates Aetna Life & Casualty (*Business Week* 1992, 142).

One source has estimated the magnitude of carpal tunnel syndrome, a debilitating hand and wrist condition that has received a lot of publicity in the past few years, at 23,000 cases a year (Pinkham 1988, 52). The affliction rate for CTD disorders may be as high as 25 percent for those people who perform motion-intensive jobs with their hands (Armstrong et al. 1986, 325). One source estimates the average treatment cost for a case of carpal tunnel syndrome at $3500 and places the cost for more severe cases involving disability claims in the $30,000–$60,000 range (Hernandez et al. 1990, 795). Another source states:

> One serious case of carpal tunnel syndrome can cost up to $250,000 after surgery, compensation, legal, administrative, and lost production costs (Hebert 1990, 5).

These hidden costs of poor product design are most frequently borne by the users and associated organizations. However, manufacturers of products alleged to have caused CTD among users have been targeted in product liability suits.

A recent case analysis identified 20 occupations at risk for CTD, including material handlers, small-part assembly workers, typists, and operating-room personnel. Reportedly, operating-room personnel were at risk for several disorders, including carpal tunnel syndrome, because of the repetitive wrist flexures and ulnar deviations required to perform common surgical tasks, such as holding a retractor (Putz-Anderson 1988, 22).

Lauren Hebert, a physical therapist and president of IMPACC (Bangor, ME), consults with industry to prevent injuries in the workforce. Based on his 15 years of clinical experience treating patients with orthopedic problems of the upper torso, he confirms that medical workers are at risk for CTDs arising from poor product design, as well as poor working habits. In fact, an estimated 15 percent of his patients have medical occupations.

> You see a lot of injuries among medical workers, particularly those working in medical laboratories. In a number of cases, poor design is a contributing factor to a CTD injury.

Hebert considers CTD a significant threat to medical workers, but one that can be readily prevented through better product design and better work habits, reinforced through product warning labels and instructions.

Symptoms and Risk Factors

Cumulative trauma disorder risk factors are not fully understood, in part because the onset of symptoms is usually slow. At first, victims may dismiss symptoms (e.g., progressive pain, muscle weakness, swelling, tingling sensations, or numbness) as a temporary condition or simply as a consequence of getting older, although many victims are young adults. Later, when the symptoms worsen, the true nature of the ailment may be diagnosed as a CTD-related tendon, nerve, or neurovascular disorder.

The consensus among medical and ergonomic experts is that such disorders usually arise from cumulative, low-level trauma as opposed to sudden and severe trauma. Because individuals with the same exposure may not react the same way, it is not possible to accurately predict who will develop CTD symptoms. However, studies have identified several risk factors. Risk factors common to handheld medical device applications include the following.

Poor Posture

Wrist deviation from a neutral position places the hand at an increasing mechanical disadvantage. As a result, poor hand posture also increases the strain on the muscles and tendons, a condition that can ultimately lead to tissue damage (see Figure 16.2). For example, to perform a syringe injection, the wrist is twisted toward the little finger (producing ulnar deviation) and the thumb is extended.

Figure 16.2. *Stressful hand positions: Ulnar Deviation—bending the wrist toward the little finger; Radial Deviation—bending the wrist toward the thumb; Pinching—opposing the flexor surface of the thumb to the index finger; Extension—bending the wrist up and back; Flexion—bending the wrist toward the palm.*

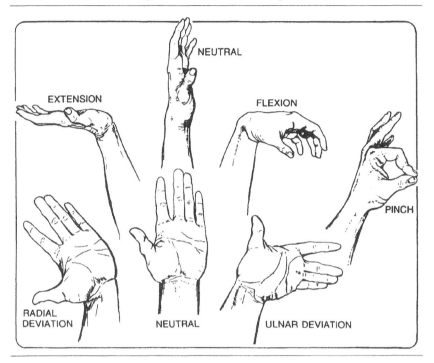

Sustained Posture

Sustained posture is also a problem (e.g., holding a retractor). According to Hebert,

> Maintaining the same posture for long periods without rest can cause injury, even if the posture is a good one, because you limit the blood flow to various body parts.

High and Sustained Muscle Exertion

As muscle force and exertion time increase, so does recovery time. This functional relationship, shown in Figure 16.3, arises

Figure 16.3. *Work and recovery guidelines for muscular exertion (MVC = maximum voluntary contractions). The longer the hold time and/or the stronger the muscular contraction, the longer the required recovery time.*

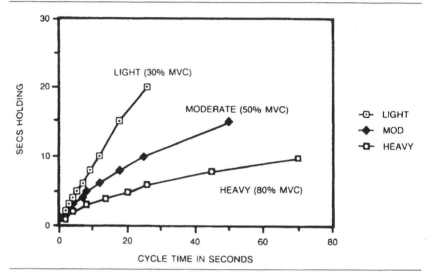

from the fact that muscles burn oxygen during exercise. The harder a muscle is used, the longer the recovery time. In the event of inadequate recovery time, oxygen deprivation (also called ischemia) can lead to muscle damage (e.g., holding a retractor for an extended period during each of several operations performed in a single day).

Mechanical Pressure

A buildup of pressure against a part of the body, such as might be produced by a tight grip on a hand tool, can compress, crush, or tear soft tissue (e.g., opening and closing various hard-edged hand tools, such as hemostats, pliers, or scissors).

Repetition

Repetitive movement can reduce the level of natural lubrication (synovial fluid) within tendon sheaths, leading to a buildup of friction. This, in turn, can produce swelling and, ultimately, the growth of fibrous tissue that constricts movement of the tendon. Repetitive motion can also set up a destructive cycle within muscles that starts with an increase in muscle tension (resistance to motion). This leads to a situation in which greater exertion is required to accomplish the same amount of work. Without rest, muscles become ischemic and tissue damage can set in (e.g., spending long hours entering patient data and clinical notes via a diagnostic workstation's keyboard).

Vibration

Vibration can restrict blood vessels, particularly in the fingers, and lead to tissue damage (e.g., using orthopedic tools to grind away bone).

Prevention Through Better Design

What role can medical device developers play in helping to protect users against various forms of CTD, particularly disorders affecting the hands? The first step is for manufacturers to become aware of the problem and its causes, particularly within the areas of a company (e.g., mechanical engineering, industrial design) responsible for designing mechanical components that make physical contact with the user. The second step is to adopt a systematic process for reviewing existing products and designs under development to determine if design changes are necessary to reduce users' exposure to CTD-related hazards.

Establishing the goal of protecting users against CTDs may actually lead to progressive design concepts that have customer appeal. For example, a product incorporating an innovative handle design may protect the user against cumulative trauma, improve worker productivity, and look attractive to users. Marketing efforts undoubtedly can make the most of such progressive designs, developing advertising claims of added user safety and comfort.

Some guidelines for designers to consider when designing hand-operated devices, adapted from *Cumulative Trauma Disorder—A Manual for Musculoskeletal Diseases of the Upper Limbs* (Putz-Anderson 1988) and other sources, are presented below.

- Design tools so they will be comfortable for people with a variety of hand sizes (i.e., 5th percentile female to 95th percentile male). Anthropometric data for the hand (e.g., *Humanscale 1-2-3*, from MIT Press, Cambridge, MA) and computer-based models of the hand (e.g., Mannequin®, from HUMANCAD, Melville, NJ) can be used in conjunction with CAD software to test the validity of designs. Good practice is to make handles at least 4 in. long with a cross-sectional diameter of 1.25 in. to 1.75 in. For tools requiring a pencil-like grip and precise movements, such as a probing device, a cross-sectional diameter of 0.2 in. to 0.5 in. is best.

- Design gripping surfaces and controls that enable users to keep their hands in a neutral, resting hand position (wrist straight with the forearm and hand partially pronated). Users will immediately perceive such grips as more comfortable. Whenever possible, design tools for both right- and left-handed use. If possible, design the grip so that users can assume alternative, comfortable postures and so that they do not have to sustain a single posture for long periods.

- When precision is not required, design objects so they can be grasped by the entire hand, rather than pinched between the thumb and fingers. Using a so-called power grip, users will strain less and be able to maintain a more secure (as much as 25 percent stronger) grip on a given device.

- Provide padding and ergonomically contoured surfaces in order to minimize mechanical stress concentrations on the skin and underlying tissues. Many computer-based workstations are equipped with padded wrist supports that keep the edges of tabletops from pressing against the tendons of the wrist. Several products, such as the familiar orange-handled Fiskar™ scissors, have contoured handles that make them popular with users.

- Select handle materials that provide a nonslip grip and protect users' hands from cold temperatures, electrical conduction, and vibration. Such handles may alleviate the need for users to wear protective gloves, which have been linked with certain CTDs.

- Provide force-assist mechanisms, such as springs and levers, to reduce the muscle exertion required to operate a device (see Figure 16.4). For example, a lightweight spring can be employed to reopen pliers after each closure.

- Analyze the range of motion of users' hands as a basis for determining the dynamic characteristics of handles and

Figure 16.4. *This tool's spring-action limits how hard users have to squeeze it.*

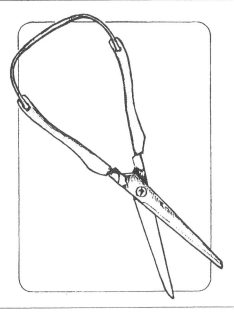

controls. For example, such analyses show that the distance between the opened handles of pliers should be in the range of 2.0–2.7 in.

- For tools with trigger controls, such as a pipetting device, a trigger at least 2 in. long will enable people to pull it using two or three fingers, thereby reducing stress on the index finger.

- Develop operational sequences that avoid frequent repetition of a movement. Analyze existing or proposed tasks to see how many times a user will perform a given physical maneuver, then study ways to reorder or eliminate maneuvers that raise the risk that users will incur.

- Balance or shield devices to minimize the amount of vibration they transmit to the user. For example, some motors and associated mounting methods produce more vibration than others. Minimum vibration can be included in the vendor selection criteria.

- Reduce the weight of objects that must be picked up or moved. Tools to be held with one hand should not weigh more than about 2 lb. Using lighter materials is an obvious way to accomplish this goal. Another possible approach for certain types of products is to relocate components from the moving portion of the product to the stationary one.

- Design heavy or awkwardly shaped objects to be grasped or lifted with two hands. Some products, for example, display labels stating the product's weight and the message: Warning: Two-person lift.

- Provide advisory instructions or visual cues on how to hold a product. One method is to label or color code the handhold positions. Another is to design grips that promote good hand posture. Grips that dictate precise finger position should be avoided, however, since hand size and finger length vary widely.

- Provide sufficient space for free forearm and hand movement so that users do not assume poor hand postures while performing tasks. Designers should investigate alternative component arrangements that afford good posture and that are also compact.

- Position work surfaces so that the forearms extend at an angle of about 90 degrees to the body, with the elbows held at one's side. The better computer workstations, those that allow users to keep their forearms in the proper position while typing, can be used as models for medical workstations. Adjustment mechanisms may be required to accommodate both large and small users.

- Avoid designs that require users to exert continuous force. Some devices, such as a hemostat or a locking pliers, incorporate mechanisms to maintain a set force.

- Use motor power in place of human power when manual tasks approach human strength limits or when tasks are performed repeatedly. The automatic inflation cuff on some noninvasive blood-pressure measurement devices reflects good design practice.

- Provide instructions to users on how to prevent CTDs. Such instructions may include diagrams of effective hand positions, a rest schedule, and special exercises.

Conclusion

Compared to other occupations (e.g., meat cutting), medical personnel do not face an extreme risk for CTD. Although an orthopedic surgeon may spend considerable time reshaping bone surfaces with a grinding tool or setting screws into bone with a specially designed screwdriver, such activities allow for diverse hand positions and moments of rest, factors that mitigate cumulative trauma. Yet there remains a low-level risk (more pronounced for users of devices such as retractors) that calls for designer awareness.

Medical device users may not develop severe cases of CTD but may experience undue muscle fatigue and joint soreness—symptoms of poor design as much as long hours of work. In such cases better device design may eliminate the mild symptoms that are precursors to more serious conditions and produce greater user satisfaction in the process. In competitive medical device markets CTD-resistant design is one way to accomplish product differentiation. It also demonstrates concern for users' needs and may reduce the likelihood of product liability claims.

References

Armstrong, T., R. G. Radwin, D. J. Hansen, et al. 1986. Repetitive trauma disorders: Job evaluation and design. *Human Factors* 28(3):325.

Hebert, L. 1990. *Living with CTD*. Bangor, ME: IMPACC.

Hernandez, J., K. Klein, V. Learned, et al. 1990. Isokinetic wrist strength of females with carpal tunnel syndrome. In *Proceedings of the Human Factors Society 34th Annual Meeting*. Santa Monica, CA: Human Factors Society.

Pinkham, J. 1988. Carpal tunnel syndrome sufferers find relief with ergonomic designs. *Occupational Health & Safety* August:52.

Putz Anderson, V. 1988. Cumulative trauma disorders. New York: Taylor & Francis.

Repetitive stress: The pain has just begun. 1992. *Business Week* June:142.

Chapter

17

Designing for Portability

Medical workers often say they wish manufacturers would improve the portability of devices used frequently or in emergency situations. Nurses, in particular, know that cumbersome devices obstruct productivity and effective patient care. Even though nurses perform their jobs stoically, they, nonetheless, have a low regard for products that strain their wrists or backs, are difficult to guide through a doorway and over strewn cabling and power cords, or require a third hand. When portability is unnecessarily compromised, they wonder: Couldn't the designers have talked to us before they designed this device? How would they like to carry it around all day?

Improving the portability of a medical device can often be achieved simply, such as by increasing the diameter of the wheels on a rolling cart, reshaping and relocating a handle to make it more ergonomic, or adding a clip to a device so that it can be attached to the patient's gown. Improving a device's portability further, however, may require that its size and weight be reduced substantially. Moving a device may require a substantial amount of mechanical work—for example, lifting and carrying a 20-lb monitor 60 ft from a storage location to a patient's room. If designers can reduce the amount of mechanical work required to transport a device, they are on the right path to improving users' assessment of it. If the amount of mechanical work cannot be reduced, steps can still be taken to maximize the user's comfort and control. Special design features, such as large, properly oriented gripping surfaces or optimal weight distribution, may lead users to judge a particular device as lighter and more portable than others, even if its overall weight and volume are the same.

User Experience

As is the case with many aspects of user-interface design, the starting point for designing a portable product is communicating with users. Users are a great source of ideas on improving a product's portability, and are generally pleased to share their ideas in an appropriate forum, such as a focus group or on-site interview. A sample of nurses' comments on the portability of medical devices is presented below.

According to an RN who works in the cardiothoracic unit of a large teaching hospital,

> I've had a lot of interaction with manufacturers developing new products. In my experience, manufacturers pay little attention to portability until the clinical trial, when problems crop up. They really should talk to us early on.

She advises manufacturers to do everything they can to maximize the portability of their devices, starting with reducing their weight and size. In addition,

> Package the device for vigorous clinical use [i.e., a lot of physical abuse]. I can't tell you how many times small devices get dropped on the floor because a patient moves his hand, or things get stacked on top of each other on the overbed table to the point that it tips over. For us, a device must be drop-proof— particularly if it's small.

Regarding larger, heavier devices, she suggests:

> Protect users from hurting themselves when they lift a device. In an emergency, people lift things that are too heavy, because they can't wait for someone to help them. We've had several nurses strain their backs lifting a venous oxygen monitor that weighs 35 to 40 lb. We ended up permanently mounting it on a

rolling cart just to keep the staff from hurting themselves. That's a short-term solution. In the future, we hope to see this monitoring capability integrated with our patient monitors so we just have to deal with a lightweight, compact, plug-in module.

Since this nurse performs a lot of patient transfers, she values medical devices that can be connected to a patient's bed, so her hands are free for pushing the bed or dealing with the patient's needs.

Anything manufacturers can do to speed up getting equipment ready for transport helps. That's a big part of the portability equation.

An unfortunate downside of making products highly portable, however, is that equipment wanders away to other units, and there is an increased risk of theft. We joke that the hospital should install car alarms on the expensive equipment so we don't lose it.

This nurse says she looks forward to the arrival of automatic equipment-tracking systems in her hospital; systems that use small transmitters that can be attached to a person's clothing or mounted on a piece of equipment and signal their location to receivers placed throughout a unit.

Another RN, who works in the pediatric unit of a community hospital, reports,

I have a real problem with our digital thermometers. We carry them around a lot, but they don't have built-in handles, straps, or clips. Therefore, they're harder to hold onto and are more likely to get dropped. I would like to be able to clip them to my clothing or attach a neckstrap to them so my hands are free for other things when I go into a patient's room.

This nurse also cites problems moving IV poles from place to place.

The way the casters are designed on certain poles makes it hard to move them smoothly across carpeted areas in our unit.

Manufacturers seeking a better understanding of users' needs and their ideas for improving the portability of existing medical devices might start by watching users perform their jobs. They might also conduct contextual interviews that include questions such as the following:

- How many times do you move the device during a typical day?

- Where do you move it to?

- How do you move it?

- Do you consider it easy to move?

- What problems do you have moving it?

- How important is making it easier to move?

- How would you improve its portability?

If a device is used by different types of people (e.g., nurses, orderlies, patients, physicians), manufacturers should talk to representatives from each group.

Another research method that can help manufacturers improve device portability is conducting a so-called endurance test, which requires users to move devices (or block models of design concepts) among various locations as normal work patterns dictate. At various stages in the workday, test participants rate the device's portability and their fatigue level on a written form, and record anecdotal information about moving the device, including compliments, complaints, and design suggestions. An analysis of the ratings may indicate a marked drop-off in user satisfaction and an increase in fatigue level after a given period. Ratings of alternative designs may show different patterns of user satisfaction, indicating that a particular design is perceived as more portable over a long period of use. Such endurance tests may last a day, a week, or a month—the length of time should depend on how important it is to make the right decision and how much test data vary.

It is best to document the findings from such user research in a report shared among design team members. Doing so will likely heighten their awareness of portability needs and may stimulate new design ideas and reinforce efforts to enhance a device's portability, even if it means substantial reengineering.

Design Factors

Device portability depends on numerous design factors, including size, weight, gripping surfaces, arrangement of accessories, ruggedness, and instructions and warnings.

Size

Making a device smaller or lighter is not necessarily the key to making it more portable. While intuition suggests that a large object is harder to move around than a smaller one, an object's proportions have a significant effect on its portability. One study of repetitive lifting shows that the maximum acceptable weight for a 22-in.-high box lifted with two hands from the floor to knuckle height dropped from 73 lb to 62 lb to 57 lb as its width increased from 14 in. to 19 in. to 29 in., respectively (Ciriello and Snook 1983, 479). Data presented in *Humanscale 4/5/6*, a set of charts that present comparative human body measurements, reflect a similar pattern for a 10-in.-high box (equipped with a briefcase-type handle at its center point) lifted with one hand from the floor to the top of a 30-in.-high table (Diffrient et al. 1981, 6–8). The maximum lifting weight for 95 percent of the male population decreased from 54 lb to 28 lb as the depth of the box increased from 6 in. to 32 in. Such data provide a good starting point for determining effective proportions for larger devices. Designers should consider conducting product-specific studies as necessary to augment such data.

Weight

The ability to lift an object depends not only on its mass but also on how it is lifted. The data presented in the Humanscale example show how the location of a device's center of gravity influences a user's lifting ability: As the center of gravity moves away from the body's centerline, a person's lifting ability decreases. Conversely, keeping a device's center of gravity close to the body improves lifting mechanics and also helps prevent lifting injuries.

General recommended limits for lifted weights have been put forward in the literature. *Humanscale 4/5/6* suggests a limit of 22 lb for women over 50 years of age, compared with 55 lb for men 20 to 35 years of age (Diffrient et al. 1981, 6–8). Furthermore, it suggests reducing these limits by 25 percent for cases when lifting is frequent. These limits illustrate how lifting ability varies according to factors such as sex and age. Therefore, designers should examine closely the intended user population for a device and set appropriate, conservative weight limits. When device weight exceeds established limits, special instructions or other design features should be introduced to discourage one-person lifts.

In some cases extra weight may facilitate portability. For example, the required configuration of a device and its accessories may make it top-heavy and it may be necessary to add ballast to the bottom to improve its stability. If this does not solve the problem, the solution may be to secure the device to something more stable (e.g., the wall, the top of a cart, or the patient bed's siderails).

Another way to improve portability is to consider the weight of a device in relation to its volume. People develop expectations for a device's weight based on its size, so a larger product that weighs the same as a smaller product may be perceived as lighter. It makes sense, therefore, for designers to attempt to discover product shapes that people perceive as lighter.

Designers should also experiment with weight distribution. Without moving the center of gravity, designers can distribute device weight in such a way that it has a substantial effect on user perceptions of weight. For example, a device that must be rotated about its axis will be easier to rotate if its weight is concentrated near the center, as opposed to near the edges.

The International Electrotechnical Commission (IEC) prescribes specific stability tests pertaining to weight distribution for the normal use and transportation of medical devices. The IEC tests call for evaluating the stability of a device when it is tilted to a predetermined angle (normally 10 degrees), with detachable parts and accessories placed in the least-stable configurations. For cart-mounted devices casters are pivoted to the most disadvantageous positions (i.e., the positions in which the cart is most likely to tip over) (IEC 601-1, 19–21).

Gripping Surfaces

Properly designed, a handle generally presents an obvious gripping surface. It should, in effect, state "lift here" or "hold here." Designers can increase a handle's visibility by means of conspicuous color coding or by placing it in a prominent position. Handles on medical devices should stand out (assuming they do not interfere with device functions), because medical workers should not have to spend undue time searching for a handle. Otherwise, they might pick up a device by a delicate part and damage it.

Designers should refrain from creating handle designs that are attractive but not obvious to the user. When designing handles, designers should consult anthropometric data sources and study hand-to-handle geometries on a wide range of hand types and associated grips (see chapter 9). Designers should also focus on how hand forces are applied to a handle. Normally, users may hold a device the way they would a suitcase—by grasping the handle with the entire hand. However, the need may arise to apply torque to the handle to rotate a device into its proper position. Thus, a pivoting handle—typical of those found on many suitcases—might be a disadvantage in terms of the stress it applies to the hand or the strength required to exert accurate control over the device.

Designers often learn the hard way that they cannot always dictate how users will hold a device. Sometimes a handle may not be the right solution for lifting or holding a device, or it may accommodate only one type of lifting or holding. Users may ignore handles altogether in favor of their own styles of lifting and carrying. Therefore, designers should consider alternative gripping surfaces that accommodate one-handed, two-handed, or multiple-person lifting and carrying.

A simple approach to this problem is to work with block models constructed from wood. Have prospective users lift and carry the models. Notice how they approach the task. Determine whether they lift the device by the handle at its top, or by using both palms to cradle the device from the bottom. Nurses moving devices and supplies among various locations in a hospital often move many things at once by stacking them. For example, medications are stacked atop bed linens, which are stacked upon a portable recording device. Designers who are aware of such practices may incorporate special gripping features into the overall packaging of a device, taking special care not to use material finishes that will be slippery when grasped by moist hands.

Arrangement of Accessories

Loose or floppy accessories can interfere with device portability. A device such as a vital-signs recorder may be festooned with various leads, a power cord, and used paper strips. When in a rush, medical workers may grab the device without taking the time to wind up the power cord, tear off the used paper strip, or bundle up and store the lead wires in order to protect them. As they rush toward their destination with recorder in hand, the cord may fall to the side and trip the user, the recorder paper may get mangled, and the lead wires may get crimped under other supplies. An improved design might include a recoiling power cord, a drawer for the leads and supplies, and a mechanism for automatically capturing paper printouts until the user removes them.

Ruggedness

As one of the nurses surveyed stated, medical devices must be designed for rigorous clinical use, which translates as continual (though unintentional) abuse. Therefore, designers should take steps to protect delicate device components. One solution is to place them in locations on the device less likely to receive direct impact, where they won't get banged against the user's body or whacked by a door jamb as the person carrying the device enters a room. Another solution is to place guards around delicate portions of the device.

Manufacturers can safely assume that portable devices will get dropped repeatedly during their useful life, so they need to design them to withstand such incidents. In fact, many products intended for medical use must pass drop tests prescribed by standards organizations such as Underwriters Laboratories, as well as specific corporate standards. Such tests determine whether a device can survive a worst-case drop from a specified height, such as the top of a mattress, without incurring significant damage.

Instructions and Warnings

Written instructions or warnings may help to ensure the safe and effective transportation of a medical device. Although manufacturers should not rely solely on such messages, they may encourage users to follow optimal transport procedures. Examples of such messages include the following:

- Lift here.
- Lift only by handles.
- Two-person lift.
- Weight = 22 lb.
- Do not tilt.
- Always hold with both hands.
- Close all drawers before moving.

Instructional messages and warnings should be developed with care so that they communicate reliably to intended users. Text may be combined with signal words (such as warning or caution) and graphics to attract attention and improve understanding (see chapter 27). Messages should be placed in positions where they will be most conspicuous to users transporting the device.

Manufacturer Experience

Larry Hoffman, a senior industrial designer with Hewlett-Packard Company (HP) (Waltham, MA), has more than 17 years experience designing medical devices such as portable patient monitors, defibrillators, and clinical information workstations. He and his colleagues place great importance on maximizing the portability of their mobile products.

> Our users, who are mostly nurses, work under considerable time pressure and have a lot on their minds. Transporting a medical device should be viewed as a necessary evil—a task we want to make as transparent as possible, letting the nurses focus on their overall patient-care goals.

Hoffman feels that portability starts with minimizing the size and weight of a given product, but recognizes the practical limits of such measures.

> In addition to minimizing bulk, we study carefully how users transport a device, and we look for ways to improve the process.

In developing HP's component transport system (Figure 17.1), which is used to monitor the condition of patients en route from the operating room to the intensive-care unit, HP's design team conducted focus groups at several hospitals, convened nurses' panels to review the design in progress, and asked prospective users to carry around weighted (about 20 lb) block models to assess user comfort.

Referring to the block model experiment, Hoffman says,

> We evaluated varying handle styles to determine which best suited user needs and preferences.

A special design challenge was providing separate but closely spaced handles enabling users to carry either the entire transport monitor or its snap-in module rack containing portions of the device's electronics.

Figure 17.1a. *Hewlett-Packard's component transport system, which is used to monitor patients traveling from a hospital's operating room to its intensive-care unit. Note that the shelf on which the monitor is placed is designed to flip up over the end of the bed.*

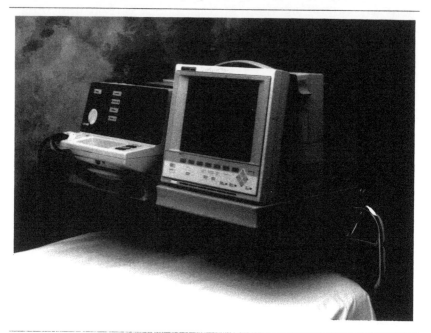

Courtesy of Hewlett-Packard Company.

We made a conscious decision to subordinate the rack's handle to the monitor's handle by designing it so that you cannot grasp it with all of your fingers. We wanted to discourage users from lifting the entire monitor by the rack handle, since it does not provide as balanced a lifting point and it strains the rack-to-monitor connection. Instead, you can only fit about three fingers in the recessed handle—more than enough to lift the relatively lightweight rack [Figure 17.2].

The monitor's larger handle is designed to stand out as the intuitively obvious place to grasp the monitor. The monitor handle is designed to fit all of the fingers under a flexible strap, and is aligned with the center of gravity of the overall device.

As an added precaution against lifting the monitor by the rack handle, HP added a warning label stating:

Caution—do not lift monitor assembly with this handle.

Figure 17.1b. *Hewlett-Packard's component transport system in action.*

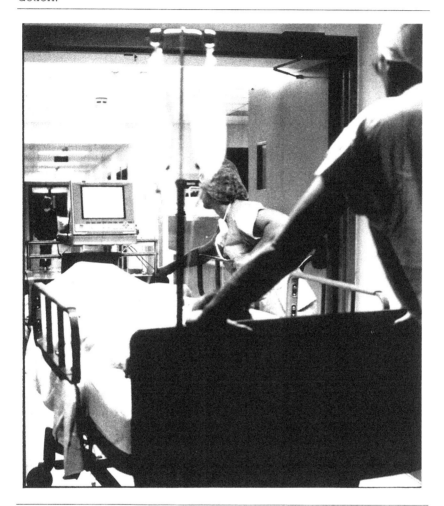

Courtesy of Hewlett-Packard Company.

To add to the product's portability, HP also designed the monitor to attach to a roll stand that sits on a specially engineered shelf that fits on a bed and flips up at the patient's feet; it also can attach to a wall mount with a quick-disconnect feature.

Hoffman estimates that a transport monitor might typically be moved four to eight times a day, allowing ample opportunity over the course of the product's life cycle for the computer-based device to get banged around or dropped. Therefore, the device

Figure 17.2. *The handle on the satellite rack of Hewlett-Packard's component transport system is designed to accommodate only a few fingers, which discourages attempts to use it to lift the entire monitor.*

Courtesy of Hewlett-Packard Company.

was subjected to considerable mechanical-abuse tests. Hewlett-Packard used test procedures derived from Underwriters Laboratories' medical and dental equipment specification, which exposed test monitors to various impact loads (UL 1976).

Hoffman has also wrestled with the task of making cart-mounted devices, such as clinical information management computers and diagnostic devices, more portable. He says that mounting a device on a cart is the appropriate solution when one or more of the following conditions apply:

- The device is too bulky or too heavy to carry around by hand.

- The device is too expensive to place several in dedicated locations.

- The device needs to be placed at a specific height or distance from the patient.

- A cart is needed for storing supplies required to use the device.

- The device is delicate and needs special protection.

To improve the portability of HP's CareVue 9000 clinical information system, which incorporates a 19-in. color monitor, microcomputer, keyboard, and trackball, Hoffman designed a cart with large, 4-in. diameter wheels to maintain stability when it is rolled over a hospital-grade power cord (about 7 mm diameter) at a speed of 3 ft/sec, and a large handlebar that gives users a range of grip positions.

> One of the most important features is the locking casters, which lock out both swivel and wheel rotation. We put locks on the front casters—those closest to the user when viewing the monitor—so that they are accessible from the normal push/pull position and when the workstation is backed up to a wall. To optimize steering and straight tracking, we pivot the front casters but not the back ones.

Conclusion

The high cost of many medical devices means they become shared resources that move frequently among several use locations, so portability is an especially important medical product design consideration. The busiest people in a hospital—nurses—typically do the moving and are sensitive to design features that contribute to or hinder portability. Their experience moving products among hospital units and departments forms a solid basis for critiquing existing designs and identifying opportunities for product improvement. Reaching out to such knowledgeable users and conducting portability design studies is likely to produce important product innovations and customer goodwill.

Goodwill develops swiftly when you make people's jobs easier. Considering that moving products around is a necessary evil (users would prefer devices to always be where they need them to be), medical workers view devices that are easy to transport as a gift. The design breakthrough for a device may be as simple as a neck or shoulder strap for a 2-lb device that frees the user's hands for other things, an extra gripping surface enabling a product to be held in several ways, or a quick-release bracket for a device that mounts on a wall or a bed rail. Even if more substantial design changes are needed to improve portability, such as reducing total product weight by 20 percent, the investment will likely pay off. Portability gains are fodder for persuasive marketing claims and produce immediate, positive reactions from users who sense their work life is becoming easier.

References

Ciriello, V., and S. Snook. 1983. A study of size, distance, height, and frequency effects on manual handling tasks. *Human Factors* 24(5):479.

Diffrient, N., A. Tilley, and J. Bardagjy. 1981. *Humanscale 4/5/6.* Cambridge, MA: MIT Press.

Medical and dental equipment, Standard UL-544. 1976. Northbrook, IL: Underwriters Laboratories.

Medical electrical equipment, Part I: General requirements for safety, IEC 601-1 (2nd ed). 1988. Geneva: International Electrotechnical Commission.

Chapter

18

Designing Packaging for Convenience and Safety

Medical device manufacturers miss an important opportunity if they treat packaging as if it were a trivial product component. After all, a well-designed package makes workers' jobs easier and contributes substantially to the overall appeal of a product. Doctors and nurses, for example, will react favorably to packages that save them time and effort. Conversely, medical workers will avoid using an otherwise well-designed product if its package creates extra work or compromises personal safety. For instance, a poorly designed package may induce nurses to spill its contents on themselves or the floor, creating potential safety hazards that may require complicated cleaning operations. Such problems can disrupt critical medical procedures, reduce worker productivity, and increase medical costs. For these reasons it makes sense for manufacturers to invest a portion of their product development budget on designing a safe, usable package.

Getting Users Involved

The task of designing a usable package begins with a solid understanding of users' needs, which means involving users in the design process. When updating existing packaging designs, manufacturers should ask representative users open-ended questions, such as the following:

- Has our package created a usability or safety problem for you?

- How could the package be made easier and safer to use?

- Would a different packaging design make using the package and its contents more appealing or satisfying?

- Manufacturers of new products should consult users in a similar fashion to define package requirements and solicit reactions to design concepts.

User feedback may be obtained from focus group discussions or from interviews conducted at the work site. Both techniques invariably lead to useful design ideas—for example, increasing the visibility of the tabs that users pull to open a package by making them a color that stands out against an opaque plastic background, or fine-tuning the tab's pulling resistance so that a package is less subject to jerky motions that can lead to spills. Most users—staff nurses in particular—will enthusiastically share their likes and dislikes about a specific design or their views on a more general design issue. The relative importance of various usability attributes can be differentiated using a technique called conjoint analysis, a systematic approach to assigning design priorities.

The most valid approach to usability testing, of course, is to involve real users (doctors, nurses, lab technicians, aides, etc.). If packages are used exclusively by operating room nurses, then the test subjects should also be operating room nurses—perhaps 10 or so. If a package will be used by people in many different types of job categories, a larger sample may be required. Five test subjects per user subpopulation may be the best compromise if there are several subpopulations and the time available for testing is limited.

Representative-use conditions, or constraints, should be established when tests are conducted. This may mean subjects wearing gloves made slippery by a given fluid, placing a package in a congested storage location that has poor lighting, or placing an unrelated device in the user's preferred hand so that he or she has to use the nonpreferred hand to grasp and open the test packages. The best approach may be to conduct tests in a clinical environment; however, a mock clinical setting may suffice.

The data from such testing may indicate that the package is going to work well for users. Conversely, test results may expose problems with a package, such as the following:

- Users require a long time to find a package when it is placed in a drawer or on a shelf with similar packages.

- Sterile contents frequently become contaminated when users open the package.

- Users require a long time to open a package.

- Users are unable to open the package in the intended manner.

- Package contents are frequently dropped or spilled.

- Users get paper cuts or scratches from the package edge.

- Users complain that a package is awkward to handle.

- Users complain that opening a package takes too much physical strength, dexterity, or mental concentration.

An analysis of such findings will generally indicate the design improvements that are necessary.

Whenever researchers undertake user studies, they must assume that usability problems stem from design shortcomings rather than from user incompetence. Usually, a case of perceived incompetence can be traced to design elements that exceed the users' capabilities under given use conditions. Therefore, designers should develop a "worst-case scenario" as a basis for design criteria. Such a situation might require an individual who has no prior experience with a package to locate or open it in an emergency situation.

Communicating Effectively

A useful strategy for package designers is to regard packaging as a communication medium. Consider the example of a nurse faced with the task of finding the correct size syringe in an operating room cabinet filled with packages of every description. Initially, the nurse's task may be one of pattern recognition, followed by a closer inspection of the package label to confirm its contents. The package's material, size, shape, and labeling communicate its identity and, depending on its design, will either facilitate or hinder the nurse's search.

One way to test users for rapid recognition and information recall is to allow them a brief glance (perhaps three seconds) at a graphic design and then ask them to identify the type of product and recall pertinent details, such as syringe size. Another approach is to show users several package designs and ask them to rank them in order of relative preference. Users should also explain the reasons for their selections. This approach facilitates a rudimentary statistical analysis of the data, which may indicate that one design is strongly preferred by users or that a hybrid design is best.

All packages should be designed so that they readily reveal their contents and other important details, such as fluid concentrations. Packages need not compete for attention, except in cases where a device has an emergency function, such as a trachea tube. Of course, such emergency devices often get taped to the wall, where they can be reached in a matter of seconds. Nurses suggest that manufacturers use large, bold labels to identify products. The labels should set off essential information from supplemental information and be placed on surfaces that are visible whether a package is in storage (e.g., in a medication cart drawer, in a cabinet, or on a supply room shelf) or being handled. Nurses feel that using color as a coding mechanism is a good idea, although they recognize the problem of establishing consistency among manufacturers.

Preferred Features

Interviews conducted on the subject of medical packaging with selected medical workers yielded both positive and negative responses. While medical workers offered more complaints than praise, in general, they felt that significant design improvements have been made over the past decade. A sample of medical workers' likes and dislikes follows.

A dermatologist's aide likes packages that serve as both a storage device and dispenser, a feature commonly used for bandages and medication.

> These types of packages help me keep the work area clean and organized, and they eliminate the need for additional dispensing devices.

A registered nurse from a major teaching hospital commented,

> I like it when things are packaged in small, plastic tubs—for example, antiseptic swabs or sponges. The tubs are easy to open, particularly when the pull tab is easy to find and grasp and the top peels off cleanly. I am concerned about the amount of plastic waste this type of packaging generates, but the convenience seems worth it.

A urologist from a community hospital said,

> It's nice when related items are packaged together. Catheter kits are a good example. The kit comes as a box wrapped in plastic, and inside are the catheter, drainage bag, betadine solution in its own tear-open package, cotton balls, forceps to hold the betadine-soaked cotton balls, syringes filled with lubricant and saline, a container for a urine sample, sterile drapes, and a tray to place items in when you are performing a procedure. Everything you need is right there, which eliminates the need for a nurse to assist with the procedure.

Several nurses cited see-through packaging as a real advantage for certain types of devices. A nurse from a medical surgical unit said,

> In some cases, you pick things out of a drawer in a big hurry. There are times when you have to cut and paste items together from several different packages to build exactly what you need. In these cases, you want to see right away if an item is in a particular package or if the package has all its parts. A see-through window or clear package lets you do this. If you can't see what's inside, you may open the package and find it's not what you want. If you've opened a sterile package by mistake, it's wasted.

As an alternative to see-through packaging, the nurse suggested,

> Print a realistic picture of the product on the package as another means of rapid identification.

Problem Packaging

A nurse from an in-patient chemotherapy unit complained that she has trouble opening a particular brand of IV fluid, which is packaged inside a second protective bag.

You're supposed to be able to rip the bags open with your fingers, but 9 times out of 10, you have to get a pair of scissors to open them. Everybody on my floor has the same problem. Finding the scissors wastes a lot of time, particularly when you consider that one patient may require five or six IV bags in a 24-hour period and we are a 29-bed unit.

The same nurse finds certain syringe packages hard to open.

The syringes come in sealed plastic containers with a membrane lid. You're supposed to be able to press your finger or thumb against the top of the package to break the seal, but it's really tough to open it that way. We've learned to turn the packages upside down and whack them against a hard surface, like the side of a table, to get them open. It's not so much of a problem once you learn the trick. However, we have patients who have to use the same syringes at home. We end up spending a lot of time teaching them the trick to opening them.

A nurse with 23 years experience in the operating room says,

It's a real challenge to open certain types of sterile gloves without contaminating them. You have to hold them in a precise way so the outside packaging doesn't touch the gloves as you rip the package open and present the gloves to the sterile field. Gloves are often wasted because they get contaminated.

This statement leads one to wonder how often the contamination of gloves and other sterile medical devices goes undetected. An operating room nurse reports similar problems opening packages of long (15 to 18 inch) catheters.

Sometimes you need another scrub person just to help open a catheter so it doesn't get contaminated. New staff have a terrible time with them.

Several nurses also complained about confusing or hard-to-read package labels. One said,

The key information on the labels is often too small and doesn't grab your attention.

Another nurse complained about the excessive amount of information on many package labels, including oversized manufacturers' logos that may hinder the identification of the product. She is aware of government requirements dictating label content, but still feels that manufacturers can develop simpler, easier-to-read labels.

Users Seek a Voice

Medical workers have well-formed thoughts on what makes a good package for medical devices. As with many areas of device design, they are eager to share their thoughts with designers, but see limited opportunity to do so. In addition to focus groups and usability tests, manufacturers could establish formal mechanisms to obtain user feedback, such as comment forms and toll-free telephone numbers that users can call to register complaints and advise manufacturers on possible design improvements. Some manufacturers establish geographically distributed panels of 10 to 15 users who are compensated to provide continuing design feedback.

The efforts that some manufacturers make to reach out to users are misdirected. One nurse advises,

> Manufacturers should take care to get input from the right people. Often, you see requests for feedback go to the director of nursing or some other administrator.

This often happens because manufacturers arrange to get user feedback through a sales or marketing representative, who works with an established contact, typically someone in management. Instead, the nurse says,

> they should reach out to the staff nurses, the people who use the products all day long and can tell them what is good and bad about the packages.

Above all, manufacturers should resist superficial methods of package evaluation, such as asking five members of their word-processing staff to try to open the latest prototypes. Basing product evaluations on such in-house evaluation is a sure way to miss potential design problems. Manufacturers should also beware of using the same in-house clinical specialists for all evaluations, since they begin to lose their objectivity after continued, close involvement with a manufacturer.

A Role for Academia

An improved understanding of packaging usability issues may come from academic research, perhaps with industry sponsorship. Hugh Lockhart, professor and associate director of the Center for Food and Pharmaceutical Packaging Research at Michigan State University (East Lansing, MI), says,

> We need to bring together existing anecdotal evidence and clinical experience with lab evaluations of packages. We don't know enough about how people interact physically with packages and what they are feeling at the time they use them.

Lockhart encourages manufacturers to perform product-specific usability studies.

> I expect that an investment in usability would give a manufacturer a competitive advantage.

Meanwhile, Lockhart would like industry groups or manufacturers to underwrite academic research. He feels that such research could lead to major improvements in packaging design.

Limited studies conducted by Lockhart have already produced useful results that defy expectations.

> We had people open several packages by pulling on a tab and then indicate which package required the least pulling force. Some people perceived significant differences. However, our lab tests showed that the forces required were about the same, which tells us that the users' preferences do not always match lab results.

Lockhart and his associates determined that people's perceptions of pulling force are tied to overall package mechanics as much as to the pull tab's resistance. Frustrated by a dearth of research on package design issues, he gives the industry only an average grade for user-interface design quality, but hopes to see improvements in the next few years if research is funded.

Material Suppliers

Medical device manufacturers frequently find that they must depend on packaging solutions developed by outside packaging suppliers, even if they operate an in-house packaging line. This situation has advantages, however. Some suppliers are willing to invest their own R&D resources to evaluate new or existing packages, particularly if they can substantially increase packaging sales as a result. Frank Arthur, in charge of healthcare marketing at American National Can (Chicago), a packaging supplier to the medical device and other industries, says,

We have a substantial R&D budget that could support viable usability evaluations. On occasion, we have taken prototypes from low-production runs into a hospital and paid nurses to give us their reactions. This helps us fit the package to the users' needs.

Arthur encourages more customers to avail themselves of services that may be provided by suppliers without cost to the medical device manufacturer.

We find ourselves doing a lot more user testing for customers in the food industry than in the medical industry, perhaps because so many medical device manufacturers prefer to do their own in-house evaluations. However, we would be eager to apply our R&D capability on their behalf.

At a minimum manufacturers who adapt packaging designs should ensure that their suppliers have verified the usability of their packages. Suppliers should be able to cite research results or be willing to undertake appropriate studies.

Conclusion

Over the course of a year, a medical worker may search for and open thousands of packages. The cumulative time spent dealing with packages may add up to several days (see Figure 18.1). Therefore, even though packages may be viewed as

Figure 18.1. *Calculation of the time a nurse might possibly spend interacting with packages over a single year.*

Hypothetical number of packages that a nurse interacts with in a day (6 patients × 5 packages):	30 packages/day
Average time spent interacting with a package (finding and handling it):	20 seconds/package
Working days in a year (assuming a standard 5-day work week):	230 days
Total time spent interacting with packages:	>38 hours/year (approx. 1 work week)

subordinate to their contents, it is worth the time and money required to design them properly. Inadequate attention to usability concerns may lead to user dissatisfaction, wasted time and materials, and potential safety hazards.

Packaging can either frustrate users or smooth the process of obtaining and using the contents effectively. An innovative package design can yield a labor savings—for example, turning a two-person operation into a task that can be handled by one. It may also increase user safety, preventing exposure to hazardous materials. Therefore, at a time when manufacturers feel the pressure to differentiate their products from those of the competition, a user-oriented package design may be just what the doctor ordered.

Chapter

19

Keyboards: Assessing User Satisfaction and Performance Capabilities

A recent syndicated cartoon depicted an airline pilot tapping away on a keyboard, which had taken the place of the airplane's control yoke. Although intended as a parody, the drawing accurately reflects the emergence of keyboards as conduits for information to advanced devices and systems. In the medical field technologists, nurses, and even physicians find themselves using keyboards on a frequent basis. Yet, either because they are not trained touch typists or because the keyboard does not facilitate touch typing, many of these medical workers use a one-finger, hunt-and-peck style. If the keyboard is the limiting factor, the barrier to touch typing may be an unconventional key layout, smaller-than-normal keys and key spacing, or inadequate tactile feedback. This raises the question: How good does a medical device keyboard need to be?

A strict advocate of product usability might assert that all keyboards should satisfy the American National Standards Institute (ANSI) guidelines for visual display terminal workstations (see Table 19.1). However, the real-world answer to the question depends on two interrelated factors: the users' satisfaction with current solutions and the quality of the competition's keyboards. Conducting user studies and performance tests can assist manufacturers in making optimal keyboard design or selection decisions.

Determining User Sensitivities

Traditional user studies, such as focus groups or contextual interviews, can reveal whether or not users are satisfied with a given keyboard. Most medical workers will be outspoken on the matter, although some may be disinclined to make negative remarks if the research is conducted by individuals closely

Table 19.1. *ANSI guidance on keyboard design.*

Criterion	Recommendation
Key nomenclature	Minimum height: 0.1 in. Contrast ratio: 3:1
Key shape	Various shapes (round, square, rectangular) meeting key-spacing requirement
Key size	Minimum strike surface: 0.47 in. (horizontal)
Key spacing	Vertical: 0.71–0.82 in. Horizontal: 0.71–0.75 in.
Key travel	Minimum/maximum: 0.06–0.24 in. Preferred: 0.08–0.16 in.
Key force	Minimum/maximum: 0.25–1.5 N Preferred: 0.5–0.6 N

Source: *American National Standard for Human Factors Engineering of Visual-Display Terminal Workstations,* ANSI/HFS 100-1988, Santa Monica, CA, Human Factors Society, 1988.

involved with the product. It is not uncommon for users to react differently to the same feature, as in the following comments:

> I like the way the rubbery keyboard protects from spills, which are a common occurrence in the operating room and intensive-care unit. A mechanical keyboard would be totally impractical because stuff would get in behind the keys where it is impossible to clean.

> The keys wobble and feel like mush when you press them, so you don't get a clear sense of whether or not you have hit the key properly. You are always afraid of making a mistake, particularly when entering laboratory test results.

Most often, the feedback will be a mix of positive and negative comments with the ratio indicating if there is a need to improve the keyboard.

The specific nature of the comments will indicate where improvements might be made. For example, complaints about keys that "wobble and feel like mush"—common for elastomeric keyboards—might be resolved by adjusting material thicknesses or modifying the bezel, which often serves to stabilize the keys. Complaints about key size and spacing might be addressed by enlarging the keyboard or compensated for by improving some other aspect of the design, such as key action. Complaints about keys that are defined only by graphics, as is the case on most membrane keyboards, might be resolved by adding surface relief, such as raised edges (embossing) or perhaps designing the key caps as plateau surfaces that stand slightly above the overall panel surface.

Determining Typing Performance

In keyboard performance testing the important usability attributes are the users' speed and accuracy, although they should also be asked about their impressions. The fact that the typing task associated with the medical device in question involves only a small amount of data entry is irrelevant because the consequence of error (such as entering the wrong medication dosage) can be severe.

Speed, accuracy, and user impressions can be explored by having test subjects perform timed typing tasks and then asking them to complete a rating or ranking sheet and to describe how they feel about the candidate keyboards. It is important to control for typing ability, making sure the abilities of the test subjects closely match those of the actual user population. This goal may call for a baseline survey of medical workers' skills.

At the start of a speed and accuracy test, subjects should assume representative use postures (e.g., sitting or standing in front of a console, holding a portable device in one hand). They may be instructed to type for a given period at a comfortable pace or, if users are likely to perform the typing task in a rush, at a fast pace "that does not result in an excessive number of errors." Subjects may correct their mistakes or leave them uncorrected. If subjects do correct their mistakes, researchers should count the number of corrections as well as the number of undetected errors, which can be done by reviewing a videotape that captures key presses and display responses. Alternatively, subjects can leave their mistakes uncorrected and all mistakes can be counted from a printout; however, the second approach may lead to an overly optimistic estimate of typing speed.

Keyboards that will be used infrequently can probably be assessed on a one-time, first-use basis. However, keyboards that will be used extensively are subject to so-called training effects, which may cause typing performance to improve asymptotically over time. In such cases researchers may ask test subjects to participate in an extended test over several weeks. Such in-depth testing helps to identify designs that may seem easy to use at first, but offer no real advantages in actual operation.

For example, human factors research shows that new users will enter numeric data equally fast and more accurately on keypads with a telephone layout as opposed to a calculator layout, but that the advantage of the telephone layout diminishes as users become trained (Greenstein and Arnaut 1987, 1450–1489). Similarly, a linear arrangement (the row of numeric keys on typewriter keyboards) may initially seem quite acceptable for entering numeric data, but is undesirable for applications requiring intensive numeric data entry, such as entering large sets of laboratory test results.

Incorporating keyboards in some devices may introduce a fundamental change in the nature of worker tasks, and some medical personnel may be concerned that typing information rather than writing it down will slow them down. Researchers can address this issue, in part, by including a writing exercise along with the typing tasks in their performance study. By timing the exercise, one can draw comparisons to typing speeds and estimate changes in throughput.

Some medical device manufacturers may not have the time, money, or motivation to conduct in-depth performance tests. In those cases a "quick and dirty" approach is better than nothing. For instance, over the course of a single day, a sample of 10 or more company employees (not members of the product development staff) can be asked to press keys on alternative keyboards and rank them simply in terms of feel.

Applying Test Results

One case in which performance testing led to product success involved a handheld device that users with diverse typing skills operate while in a standing position, cradling the terminal in one arm while typing numerics and short passages of text with the other. Early anthropometric studies (see chapter 9) showed that a full-size keyboard would make the product too wide to be held comfortably by people with shorter-than-average arms, so the question arose: How small can the keyboard become before typing performance is substantially degraded?

The manufacturer answered the question by conducting a throughput experiment with keyboards that incorporated various key spacings, ranging from $1/2$ in. to $3/4$ in., and self-assessed novice, intermediate, and advanced typists (Dumas et al. 1987, 585–589). Results indicated that if key spacing was reduced to about 0.70 in. (slightly lower than the minimum spacing recommended by ANSI), users could still achieve acceptable typing performance levels. By adopting such spacing, the manufacturer saved about $3/4$ in. in total keyboard width. Further reductions were made by eliminating unnecessary keys (e.g., those for punctuation marks) and by reducing the width of the normally larger keys, such as the shift key. The end result of downsizing the keyboard was a device that most users consider comfortable to hold.

Designers' Viewpoints

Discussing his own experiences with keyboard design challenges, Peter Rhoads, a design manager with Hewlett-Packard Company's Imaging System Division (Andover, MA), stressed that

> the process of keyboard design is fraught with trade-offs. You are trying to fit seven pounds of grass seed in a five-pound bag . . . . For example, we went to a smaller-than-standard

> keyboard on our latest product [the Sonos ultrasound imaging
> system] because of space limitations.

There is not very much typing involved in using the system, he
explained, adding that

> most of our users are going to hunt and peck for keys. There-
> fore, they are going to be more tolerant of a keyboard with
> compromises, compared with someone who is doing word
> processing.

Rhoads would prefer to take an iterative approach to keyboard design, using tooled prototypes, but feels economic barriers prevent such an approach. Instead, he seeks informal
reactions to existing keyboards, reviews the literature on keyboard design, and draws on the expert judgment of his staff,
which includes a human factors specialist. He once considered
stereolithography as the means to evaluate designs on an iterative basis, but,

> the parts come out feeling like sugar candy, which makes them
> incompatible with an evaluation of keyboard feel.

Nonetheless, Rhoads sees the value of more-exhaustive keyboard evaluations for devices where the design issues are complex and keyboard performance is important to the product's
success.

Gary Shepard, an industrial designer with Hewlett-Packard
Company's Cardiology Business Unit (McMinnville, OR), also
understands the benefits of research. When developing the keyboard for the PageWriter XLi cardiograph (Figure 19.1), he
recounts,

> We conducted field research on products incorporating key-
> boards that are used in hospitals. Among other things, we
> found that the keyboards were filthy. Right away, we decided
> to look for new ways to keep keyboards clean without resorting
> to membranes, since they lacked the tactile qualities we were
> after.

The solution to this challenge was a mechanical assembly of
keys that mounts on top of a separate elastomeric keypad. Users
can snap out the key assembly and immerse it in cleaning solution, and the underlying elastomeric keypad can then be wiped
clean without contaminating the keyboard electronics. Shepard

Figure 19.1. *Hewlett-Packard's PageWriter XLi cardiograph with snap-out keyboard.*

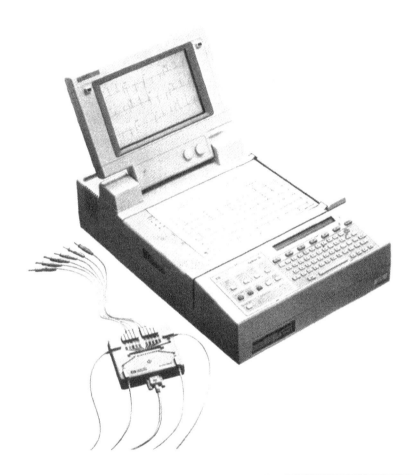

Courtesy of Hewlett-Packard Company.

reports that the keyboard has been well received by users, particularly because the keys provide good tactile feedback, even though they are spaced only 0.675 in. apart in order to make room for other cardiograph controls.

Conclusion

Healthcare technicians and providers will learn to use whatever keyboard is included on a medical device. However, such users are becoming increasingly sensitive to keyboard quality by virtue of their experience with microcomputers. Although a device keyboard they come to perceive as inferior may not necessarily interfere with their successful completion of a task, it may color their experience of using the product. Therefore, device designers should take care to ensure that the keyboards they choose will both satisfy users and perform well. Otherwise, the keyboard may emerge as a product's Achilles' heel, leading customers to prefer a competitor's offering.

References

Dumas, J., L. Hoffman, and M. Wiklund. 1987. Optimizing a portable terminal keyboard for combined one-handed and two-handed use. In *Proceedings of the Human Factors Society 31st Annual Meeting.* Santa Monica, CA: Human Factors Society.

Greenstein, J., and L. Arnaut. 1987. Human factors aspects of manual computer input devices. In *Handbook of Human Factors*, edited by G. Salvendy. New York: John Wiley.

Chapter

20

Choosing an Effective Pointing Device

Consider the trend in user-product interactions. Diagnostic procedures such as computed tomography scanning and ultrasound imaging are dramatic examples of the shift toward computer use in modern medical procedures. Instead of conducting exploratory surgeries, doctors often direct a computer to generate the video images necessary to make a diagnosis. Ultimately, the patient's medical condition is revealed by the movement of a cursor instead of a scalpel.

Because product functions are now embedded in software, many medical devices require computer interaction skills, such as navigating through menu structures, selecting control options, inputting data into fields, scrolling across graphic displays, and moving objects rather than turning knobs, moving levers, and pressing buttons.

Hence, pointing devices, also called cursor-control devices, are needed. Hospitals are now populated by mice, trackballs, joysticks, touch screens, light pens, graphic tablets, so-called trim knobs, and other miscellaneous pointing devices, each offering advantages and disadvantages from a human factors standpoint. In order for medical device manufacturers to make the right choices, they must at least understand these trade-offs, and they may well have to conduct human performance testing.

Many Options

A wide variety of pointing devices are available, which is fortunate because a device that works best for target selection (pointing to a specific screen location), for example, may not be well suited to text editing or free-hand style drawing. Whether a pointing device is the right choice for an application depends on input resolution (how accurately one must point to screen locations) and on whether tasks involve special control modes, such as selecting an object and then moving it. Physical constraints, such as limits on available horizontal surface area (a device's "footprint"), may lead designers to favor one device over another. For example, clinical environments such as the intensive-care unit generally are congested places that make it difficult to use cursor-control devices with a relatively large footprint, such as the roughly 9 inch × 9 inch area need to use a mouse effectively.

Also, some devices are easy to use at first, while others are initially somewhat difficult to use, but become easy once the user gets accustomed to them. This last consideration can be made clear using the metaphor of learning to ride a bicycle, an experience that can be arduous and bruising, can take a relatively long time, and may require special learning tools (training wheels). On first try a cyclist would likely rate the device as "very difficult to use." However, once a cyclist becomes proficient, he or she considers bicycle riding to be second nature, a task requiring little conscious effort. The trackball and mouse compare aptly to the bicycle in this regard. Initially difficult to manipulate accurately, each device becomes a natural extension of the hand after a few hours of use. In fact, many experienced mouse or trackball users would likely consider a touch screen to be a slow, annoying method of human-computer interaction, largely because of the lack of kinesthetic feedback and arm fatigue. Furthermore, touch screens typically require screen designers to place less information on a given screen, as compared to one calling for mouse or trackball interactions, to accommodate large touch targets (about 0.5 inches in diameter is the practical minimum size, although smaller targets are workable given certain affordances, such as a "lift-off" actuation scheme).

Because the mouse and trackball can be used for long periods without creating fatigue, they make appropriate choices for medical products requiring user interactions on the order of an hour or more per day. By contrast, because they are so easy to use, touch screens are an appropriate choice for applications requiring structured user interactions (not including text entry), interactions performed under stress and with divided attention, and intuitive use by first-time or infrequent users.

Controlling the cursor by means of arrow keys can be a particularly slow solution for most applications. However, arrow keys work well in applications requiring discrete and predictable movement among options, as though one were using the tab key within a word processing application. In place of arrow keys, many manufacturers have turned to trim knobs, which allow users to move rapidly among appropriate on-screen selections and fields. For example, Marquette Electronics, Inc. (Milwaukee), uses trim knobs as the principal means of interacting with their patient monitoring products, a commitment that seems to have paid off in high user acceptance.

Cursor-control devices such as the joystick, touch tablet, touch pad, light pen, and isopoint are used on a much more limited basis, but may be optimal for special purposes.

Conducting Performance Tests

Medical product developers frequently find themselves choosing from among several types of pointing devices. Sometimes, a review of the human factors literature will clarify the selection decision. More often than not, however, existing scientific data may not apply to the design problem at hand, suggesting the need for specially tailored performance tests.

In one approach to testing, subjects perform realistic tasks with a user-interface prototype or a working model, enabling researchers to obtain accurate impressions of how a pointing device will actually work in a particular application. However, this approach may make it difficult to determine why errors occur and may obscure how specific variables, such as direction of cursor movement, influence overall performance.

In a second approach subjects perform discrete cursor-movement tasks using custom-built testing software or only limited portions of an existing user interface. For example, subjects can make vertical cursor movements from one target (data cell) to another on a Lotus 1-2-3® spreadsheet. This approach helps researchers determine whether test subjects can perform specific tasks easily, but it does not enable them to analyze how well subjects perform in real-life situations.

Following either approach, prototyping tools developed for use on Macintosh or PC-compatible computers can provide a cost-effective and relatively fast means for creating the necessary test software, particularly when working models are not available for testing (see chapters 21 and 22). In fact, more time can be spent planning such tests and finding test subjects than developing the test software, which can take only a matter of hours or days.

A comparative test of four pointing devices, including all of the preparation, testing, and reporting steps, might take two researchers up to four weeks to complete, although the total time required for testing will vary according to the number of test subjects. Depending on the experimental design, an unbiased test may require more than 20 subjects with appropriate clinical backgrounds. Using larger subject samples can improve researchers' confidence level, but such an approach may not be cost beneficial. Often, it is best to plan a limited test and determine the significance of the test data before running additional subjects.

The first step for a manufacturer to take in developing a test to choose a pointing device is to consider the nature of the required pointing tasks. Users of ultrasound imaging devices, for example, frequently point to a spot on a displayed image (i.e., they acquire a target). This action could be tested using a test item similar to that shown in Figure 20.1, which calls for subjects to move a cursor over each of several rectangular targets and select them using the appropriate method (e.g., a button press on the mouse). Another imaging task involves drawing boundaries around objects of interest, such as a circle or ellipse around a fetus's head. This might be tested using a test item similar to that shown in Figure 20.2, which requires subjects to draw circles concentrically around existing circles. Human factors literature provides additional examples of such tests (Epps 1987, 442–446); developers can create test items to match their test goals.

Figure 20.1. *Test item for assessing target acquisition.*

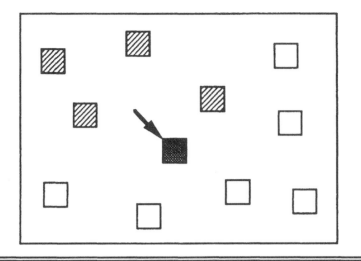

Figure 20.2. *Test item for assessing drawing.*

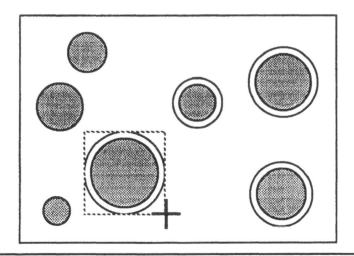

An Expert's Viewpoint

Joel Greenstein, Ph.D., associate professor of industrial engineering at Clemson University (Clemson, SC), has conducted extensive research on pointing-device effectiveness and has written several authoritative articles on the topic.

> The design practitioner is probably best served by having test subjects perform realistic tasks using the candidate pointing devices. It is not [the test subjects'] job to fill in the blanks in the scientific data base on pointing devices, which would require a more constrained, scientific experiment. Practitioners need to look at the integrated performance of a pointing device in their application via user testing. Although information on performance attributes can be found in the literature, what is missing in the literature is information about how the pointing device will work for the designer's specific application.

Despite his endorsement of real-world testing, Greenstein still sees value in tests that look at one variable at a time. He recommends this approach to those developing a custom pointing device or specifying certain performance characteristics, such as device gain or resistance to motion. In general, Greenstein advises:

> When choosing a pointing device, make the decision within the framework of designing for usability, as described by Gould and others [Gould 1988, 757–789]. Do not base a selection decision simply on facts and guidelines about pointing devices [Greenstein and Arnaut 1988, 495–519]. You have to study who will use the device and the tasks they are going to perform with it. Then you can begin to set behavioral objectives that give you a stronger basis for evaluating and selecting the best device.

Greenstein's research on the performance of pointing devices began in the mid-1980s, when he conducted experiments for the U.S. Navy. The Navy was concerned with selecting the optimal device for human-computer systems installed on ships.

> For military systems, you place a lot of emphasis on the objective performance of a pointing device. You are most concerned about effecting action quickly and without error, since the consequence of slow response and errors can be loss of life. The subjective issues [whether users like a device] are of less concern, since the user is not presented with the option of not using the system.

Although loss of life could also be a consequence of ineffective human-computer interactions with medical products, Greenstein considers subjective issues to be of greater importance for medical applications.

> With medical products, the manufacturer has to be concerned with users' satisfaction with the product. If users do not like a given pointing device, they will seek ways to get their job done without it.

Comparing the relative importance of subjective versus objective issues, Greenstein says:

> It would be difficult to dismiss a pointing device that was most preferred, even if its objective performance were somewhat lower than that of an alternative device. However, you would have to look at the consequences of slower task times and associated errors to make sure that they do not create major problems. Users are going to make errors with a pointing device, so you should design to ensure that users can detect and correct them. It would be unusual to find a pointing device that produced rapid, error-free performance that was not well received by users.

Based on research and applied design experience, Greenstein (and coauthor Lynn Arnaut) developed a chart that delineates the advantages and disadvantages of various pointing devices (see Table 20.1). Greenstein recommends that people use the chart as a starting point for determining viable options and not as the exclusive basis for selecting a pointing device.

One Manufacturer's Experience

Hewlett-Packard Company manufactures several computer-based medical products requiring extensive cursor-control activity, including several patient monitors, diagnostic ultrasound imaging systems, a portable clinical information management workstation, and an arrhythmia information management workstation. One patient monitor uses a beam-type touch screen, another uses cursor keys arranged in a diamond pattern, and the other products employ a trackball as the pointing device. The company's latest arrhythmia information product

Table 20.1. *Advantages and disadvantages of pointing devices.*

	Touch Screen	Light Pen	Graphic Tablet	Mouse	Track-ball	Joy-stick
Eye-hand coordination	+	+	0	0	0	0
Unobstructed view of display	–	–	+	+	+	+
Ability to attend to display	+	0	+	0	+	+
Freedom from parallax problems	–	–	+	+	+	+
Input resolution capability	–	–	+	+	+	+
Flexibility of placement within workplace	–	–	0	0	+	+
Minimal space requirements	+	+	–	–	+	+
Minimal training requirements	+	0	0	0	0	0
Comfort in extended use	–	–	0	0	+	+
Absolute mode capability	+	+	+	–	–	0
Relative mode cabaility	–	–	+	+	+	0
Capability to emulate other devices	–	–	+	–	–	–
Suitability for:						
pointing	+	+	+	+	+	–
rapid pointing	+	+	0	0	0	–
pointing with confirmation	–	0	0	+	0	–
drawing	–	–	+	0	–	–
tracing	–	–	+	–	–	–
continuous tracking, slow targets	0	0	+	+	+	–
continuous tracking, fast targets	–	–	0	0	0	+
alphanumeric data entry	–	–	–	–	–	–

Key to symbols: + = positive, 0 = neutral, – = negative.

Source: *Handbook of Human-Computer Interaction* (see References).

ships with both a mouse and a trackball, providing users with a choice. George Adleman, a human factors specialist in Hewlett-Packard's Waltham Division (Waltham, MA), says,

> We pay close attention to human factors as well as to technical and manufacturing issues. However, we also have to be concerned about supporting the product in the field. You don't want to make pointing devices that will present users with reliability and maintenance problems down the line.

According to Adleman, the choice of pointing device for Hewlett-Packard's clinical information management workstation (CareVue 9000) was a significant issue. Early in the development process designers had envisioned using a touch screen in conjunction with a keyboard as the means to manipulate clinical information (e.g., vital signs data) on a large spreadsheet. They reasoned that touch would be an intuitive interface for occasional users, such as circulating physicians, while it would afford good usability for more frequent users. However, company engineers came to view touch as a high-risk option because of the limited availability of large touch screens and because of typical concerns about touch screens: parallax and reduced optical clarity (chronic with overlay technologies) and surface contamination (primarily fingerprints).

In place of a touch screen, the company considered a trackball, mouse, and isopoint. According to Adleman,

> We tested these devices along with touch as a baseline.

The test involved 12 subjects (nurses working in critical-care environments) who performed discrete cursor movements (toggling horizontally, vertically, and diagonally between targets) on sample screens excerpted from the user-interface prototype. The prototype was developed in Dan Bricklin's Demo II (Sage Software, Beaverton, OR) and ran on a Hewlett-Packard Vectra computer (PC-compatible). Hewlett-Packard researchers collected time and error data and surveyed the nurses' preferences.

> We found touch to be twice as fast as the other options, and it was the preferred option.

Determining touch to be infeasible for CareVue 9000 for the reasons stated previously, Hewlett-Packard was left to choose among technologies affording somewhat slower input rates.

> We found the performance of the mouse, the trackball, and isopoint to be roughly the same. We ultimately ruled out using a mouse because it required too much space on a portable

product. Also, it was not substantially better than either the trackball or isopoint. We finally chose trackball over isopoint because it was a more established technology and it afforded performance levels equivalent to a mouse.

Hewlett-Packard's Full Disclosure System, an arrhythmia information management product, presents a different set of pointing device requirements (Figure 20.3). The user interface, which has a look consistent with Hewlett-Packard's OSF/Motif® (a prominent style of user interface), calls for higher resolution pointing because the selectable targets are too small for finger-tip selection. For instance, to review EKG waveforms from an hour before, the user presses a left arrow key icon to scroll back in time to previous waveforms. Alternatively, the user can drag a selector icon along a timeline. The manufacturer's decision to ship both a mouse and a trackball with the product seems at once smart and expedient, because it eliminates the need for Hewlett-Packard to choose a device for a group of users who may already have developed a preference.

Figure 20.3. *Hewlett-Packard's Full Disclosure System.*

Courtesy of Hewlett-Packard Company.

Conclusion

In today's medical settings clinicians interact with microprocessor-based technologies on a continual basis. Such technologies are often delivered with the promise of labor savings and improved ease of use. Making good on this promise, however, depends in large part on providing users with an efficient method of interaction (i.e., an effective pointing device). The variety of available pointing devices, extensive data on device performance, and efficient methods of human performance evaluation should empower manufacturers to make the right choices.

References

Epps, B. 1987. A comparison of cursor control devices on a graphics editing task. In *Proceedings of the Human Factors Society 31st Annual Meeting*. Santa Monica, Human Factors Society.

Gould, J. 1988. How to design usable systems. In *Handbook of human-computer interaction*, edited by M. Helander. Amsterdam: Elsevier Science Publishers BV.

Greenstein, J., and L. Arnaut. 1988. Input devices. In *Handbook of human-computer interaction*, edited by M. Helander. Amsterdam: Elsevier Science Publishers BV.

Prototyping the User Interface

Chapter

21

Prototyping the User Interface

How would you improve the design of a product if you had to do it over again? We frequently ask this question of ourselves and others about a completed product, particularly its user interface. In fact, shortcomings with the user interface—that is, those portions of a product with which people interact, such as controls and displays—may seem obvious once the finished product has actually been used. Shortcomings are less obvious when a design exists only on paper or as a working model that lacks the final product's appearance and interactive characteristics. Designers of user interfaces thus often end up with a wish list of design changes they would like to have made.

For example, the designer of a noninvasive blood-pressure monitor might say,

> We should have selected a larger graphical LCD [liquid crystal display] instead of the small, character-type display that we used. The graphical display would have allowed us to draw bar graphs in place of data tables to show trend information. With a larger display, we could have made the numerical readouts larger, spread them out more for improved readability, and provided more-meaningful labels. The added cost

would have been worth it because we get the most customer complaints about the way we display information.

Or the designer of an airway-gas heater and humidifier might say,

Our device requires the user to press two keys simultaneously for certain tasks such as silencing an alarm. Most inexperienced users fail to realize that they have to press two keys. They finally give up trying to figure out how to turn off the alarm with the controls and, instead, turn the whole device off. A dedicated silence key or better labeling would probably solve the problem.

Fortunately, user-interface prototyping tools are available that enable a designer to simulate a final product's appearance and feel before detailed engineering work begins. This prototyping is done with a computer, usually a PC-compatible or Macintosh unit, but sometimes with more powerful computer workstations. Using appropriate prototyping software, the designer can draw, electronically, an image of the physical product and show its response to such user inputs as pressing a button, turning a knob, moving a lever, touching a sensitive screen, plugging in a cable, or giving a voice command. The prototype's response to such user input might be to make a sound (e.g., emit two beeps), change the text or graphics of a computer display (e.g., show revised numerical values), or change the position of a product's component (e.g., open a battery cover).

If the designer has not worked with user-interface prototyping tools before, that individual probably will think of a prototype as a near-production, handmade, mechanical model of a product. In fact, product-development teams do routinely build hardware-based prototypes before finalizing a specific design. This kind of prototype, which is viewed as being essential to design validation, usually is built late in the design process and then evaluated by clinical trial. User-interface prototypes, however, are quite different because they facilitate iterative design early in the design process, enabling usability to be engineered into a product. Accordingly, the balance of this chapter discusses how a prototype works, getting started, deciding how much effort should be invested in such design, who should create the prototype, and related benefits.

How a Prototype Works

A user-interface prototype functions much like an advanced 35-mm slide show. In other words, by showing several slides in a row, the designer can illustrate user inputs and the product responses to those inputs. For example, in a hypothetical slide show composed of three slides, the first shows a product—in this case a noninvasive blood-pressure monitor. Assuming that the monitor's start button has just been pressed, the projectionist would advance the slide carousel to the second slide, which would show a new message on the device's display, an indication that the pressure cuff is inflating. Once the hypothetical measurement has been completed, the third slide might show the system display of readings of systolic, diastolic, and mean blood pressures.

In such a slide show, advancing from slide to slide is accomplished with a hand control and in a fixed sequence. With a computer-based prototype, moving from image to image is accomplished by direct user inputs to the computer. Within the prototyping software, sets of instructions, or "scripts," are evoked by user inputs, which direct a change to the displayed image (Figure 21.1). In the case of more complex prototypes, dynamic effects can be accomplished without changing the "slide."

Once a prototype has been completed, the designer can evaluate the user interface more effectively. In particular, those individuals who might actually use the product can evaluate it for usability. Early in the design process, a sample of 5 to 15 people generally is sufficient to provide the designer with useful feedback. One recent study suggests that 80 percent of usability problems can be identified using a total of 6 subjects, and 90 percent with 10 subjects (Virzi 1990, 291–294). The number of people is less important than having the right people—that is, people who are typical of actual product users and not the designer's colleagues in the next cubicle.

Professionals who do not routinely work with computers (as well as others who do) often are amazed to find that a prototype really works. Imagine sitting in front of a computer that displays what seems to be a digital alarm clock and being told to adjust the alarm setting by pressing the buttons on the screen. First-time subjects may be intimidated by the experience and thus reserved in their initial interactions with the prototype. However,

Figure 21.1. *Prototyping image of a hypothetical gas analyzer.*

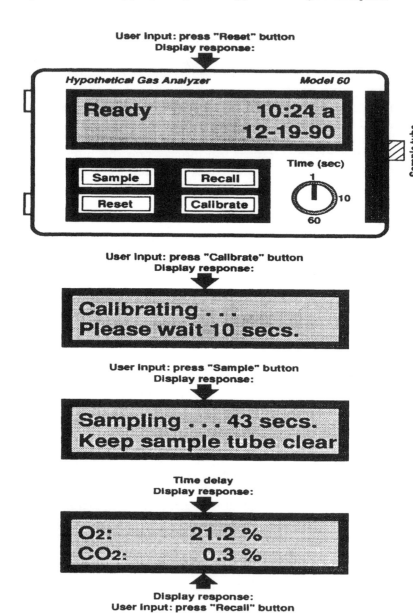

most people become quite comfortable with prototypes after just a few minutes of use. The product starts to seem authentic and the interactions become more natural.

To facilitate the evaluation of a prototype and to reduce the intimidation factor, the designer may want to have subjects perform a get-acquainted exercise that uses a simple prototype (Figure 21.2). During such an exercise people can practice button pressing and observing display responses to their inputs.

Making the experience of pressing buttons (or other types of controls) seem realistic is a major design challenge to a prototype's overall effectiveness. The button-pressing exercise may be handled in several ways. For example, the prototype can be presented on a touch-screen display and subjects need only to touch the image of a button. This action evokes an associated software-based instruction set that, in turn, changes the prototype's appearance. In another approach a subject uses a control device such as a trackball or mouse to move a cursor to the on-screen button; a separate button on the control device then simulates button pressing on the prototyped product (Figure 21.3).

Figure 21.2. *A practice screen used to help test participants feel more comfortable interacting with a prototype.*

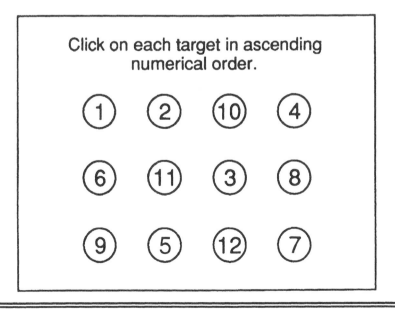

Figure 21.3. *Various prototype designs for touch screen button simulation.*

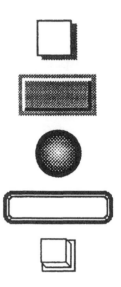

For many applications a touch screen is the preferred interaction mechanism because actions and responses do not have to be translated through a cursor-control device. The button-pressing experience can be refined further by including such audible feedback as clicks and beeps (synthesized or digitized) when a button is pressed. Special digital recording devices can make such sounds as realistic as necessary. For additional visual feedback the on-screen button can be made to move when pressed, much like the keys on a keyboard.

Getting Started

To begin prototyping, the designer should select a prototyping tool that is relatively friendly to new users, is compatible with the designer's existing computer hardware, will produce a prototype of the desired fidelity and appearance, and is supported by good user documentation and associated textbooks, tutorials, workshops, and other forms of assistance. Dan Bricklin's Demo II (Sage Software), Hypercard® (Apple Computer), and Super-Card™ (Allegiant Technologies) are three popular prototyping tools that are helpful to new users. Demo II runs on PC-compatible computers while Hypercard® and SuperCard™ run on Macintosh® computers. Hypercard® is provided on all new Macintoshes and may be the easiest of these programs to learn. All three prototyping tools enable the designer to draw black-and-white and color text and graphics on-screen. Toolbook (Asymmetrix), Visual Basic™ (Microsoft), and Director (Macromind) are also popular prototyping tools for use on a PC, although they may take new users a bit more time to become proficient with them. Another tool called Altia® Design (Altia), that once ran only on workstations but now is available for use on a high-end PC, is also gaining favor among medical manufacturers because of its powerful features.

Nonprogrammers may learn faster with the Macintosh-based prototyping tools because of the intuitiveness of the Macintosh-style user interface. Hypercard® offers an excellent introduction to prototyping because it is simple to use and features extensive on-line help; in addition, several good textbooks are available on the subject. From a user's standpoint, Super-Card™ might be considered a close cousin of Hypercard® because it uses the same general approach to creating display images and writing software instructions.

Many other prototyping tools on the market are worth a close look. Contact software retailers for information about available products.

How Much Effort Should You Invest?

Designers who use prototyping tools must decide how accurate the image of the physical product needs to be (Figure 21.4). Sometimes the product's overall appearance has yet to be established, and the image's accuracy is less important than the overall style of user-product interactions. In such cases, rectangles can be used to represent product enclosures and circles can represent knobs. At other times, industrial design drawings or earlier versions of a product may visually guide the designer to choose between developing an accurate image of the product and simplifying the image for expediency. However, an accurate image heightens the sense of realism and improves the quality of user feedback during evaluations. If the designer has a color scanner, design renderings or photographs can be used to create the electronic image on which the prototype will be based. The aforementioned prototyping packages provide the

Figure 21.4. *(a) Simplified, low-fidelity image of an LCD display; (b) higher-fidelity image of an LCD display. Note that LCD characters are shown more realistically as segmented characters.*

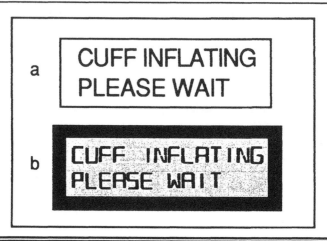

mechanisms for importing a scanned image. Such an image gives the designer the best image with the least amount of work.

A designer can spend weeks or months building a high-fidelity prototype, perhaps one that includes more than 100 separate images that illustrate various product functions. For example, a patient-monitor prototype might allow users to set up monitor functions by using various menus. The prototyping process can be sped up if the design is firmly established, although prototyping's real power comes with its use as a design tool. Designers can experiment with various user-interface design alternatives and obtain feedback before settling on a final design. In fact, designing alternatives is so easy that designers can become carried away with the process. Consequently, designers must keep their focus on the major design issues rather than on the prototype's visual impressiveness. Sometimes the guiding concept, as well as substantial development time, can be lost to the process of endlessly massaging screen artwork and special effects. A prototyping schedule will help to prevent inefficient prototyping.

Low-fidelity prototypes can be created in just days, if not hours. This fact has popularized the term *rapid prototyping*. Whether the process actually is rapid is only a matter of perspective. One can compare the prototyping effort's time with that required to build a hardware-based model and then decide whether the prototyping process is rapid.

A designer might ask,

> Why should I build a user-interface prototype when I could build a hardware-based model within a matter of weeks?

The answer depends on the type of product and previous design experience. If user-product interactions are largely physical—for example, a patient and a syringe—an electronic prototype is less useful than a working model. But if many of the interactions are visual and involve computer input and output, an electronic prototype can be quite useful and perhaps more practical to build and evaluate. A hardware-based model may show the same interactions as a user-interface prototype; however, making changes iteratively to the interface based on user feedback may be more difficult or impractical because of cost or time. On the other hand, modifying a user-interface prototype from user feedback is much easier. Prototyping makes design modifications easier in the same way that word processors make text editing easier.

Eric Smith, a user-interface specialist with Datex Medical Instrumentation (Tewksbury, MA), describes the benefits of such user-interface prototypes as follows:

> If you are trying to change a part of the user interface for improved usability, a prototype stands the best chance of illustrating the improvement and convincing people on the design team that the change should be made. A prototype lets me illustrate ideas more explicitly than can be done in words.

Smith uses SuperCard™ to develop design alternatives for patient monitors; he then shows these variations to physicians and nurses to obtain their feedback.

> Prototyping takes an up-front investment in user-interface design that adds initially to development costs. The return on the investment is a more-optimal design achieved through more-extensive design iteration and user involvement.

Who Should Create the Prototype?

Early in the design process, a company must determine which individuals should build a user-interface prototype and what skills are required to accomplish this task. In some medical companies human factors specialists have taken the lead in prototyping work, designing user interfaces that accommodate users' needs. Building prototypes is a natural extension of their work.

For companies that have no human factors specialists, other alternatives are available. Industrial designers are well suited to take the lead role in prototyping. The industrial designer's sensitivity to user-interaction issues and a well-developed sense for product appearance are real assets in developing quality prototypes. Programmers also can contribute to the prototyping effort because they are adept at learning the ways in which prototyping tools work and at handling the more-complicated aspects of prototype programming. In other words, people with varying technical backgrounds can make good prototypers and prototyping can be a team effort.

When prototyping begins early in the overall design process, the designer has the opportunity to evaluate and improve a user interface that already has the look and feel of a real product. Design problems can be detected and addressed without severely compromising product cost and schedule. Designers are free to make the user interface as good as it can be. End users are afforded the opportunity to participate in the design process and influence the important design decisions. When several competing designs are prototyped, end users can state their preferences for a specific design based on realistic experience. As a result, the designer reduces the risk of marketing a product that has significant usability problems.

Paul Weil, a product designer with Hill-Rom (Batesville, IN), has also used SuperCard™ to design the control and display features of an advanced hospital bed. He is convinced that user-interface prototyping leads to better user interfaces.

> A prototype will help you produce a better design because you are not locked into a certain design the way you are with a hardware model. You can quickly generate different designs . . . look at different control layouts and uses of color and labels. We can take a prototype to our customers and get their feedback before committing to a specific design.

Additional Benefits

The development of user-interface prototypes also brings other benefits. For example, the designer can show a prototype to potential customers before the real product is available—a potentially advantageous situation in certain marketing situations. A designer also can use the prototype to train sales personnel in how the product works. Technical writers who are responsible for product documentation (including the user manual) can work with the prototype as they write documentation, identifying ways to improve not only the user interface, but also the integration of documentation and product. The user-interface prototype also can be used to specify hardware and software, augmenting or supplanting a text specification. Of course, these additional benefits are predicated on the existence of a refined prototype.

To conclude, prototyping affords a designer the chance to explore user-interaction issues early in the design process—that is, when smart decisions that benefit the end user still can be made. Those who try prototyping probably will find that it becomes an essential part of their user-interface design process.

Recommended Readings

The following articles provide additional information about user-interface prototyping.

Metz, S., R. M. Richardson, and M. Nasiruddin. 1987. Rapid software for prototyping user interfaces. In *Proceedings of the Human Factors Society 31st Annual Meeting*. Santa Monica, CA: Human Factors Society.

Miller, D. P. 1988. Instant prototyping using HyperCard on the Macintosh. In *Proceedings of the Human Factors Society 32rd Annual Meeting*. Santa Monica, CA: Human Factors Society.

Van Kirk, D. 1989. All-purpose demo," *PC/Computing* 2(11):179.

Virzi, R. A. 1989. What can you learn from a low-fidelity prototype? In *Proceedings of the Human Factors Society 33rd Annual Meeting*. Santa Monica, CA: Human Factors Society.

Reference

Virzi, R. A. 1990. Streamlining the design process: Running fewer subjects. In *Proceedings of the Human Factors Society 34th Annual Meeting*. Santa Monica, CA: Human Factors Society.

Chapter

22

User-Interface Prototypes: How Realistic Should They Be?

As discussed in the previous chapter, building a dynamic, computer-based prototype of a device's user interfaces five years ago was a novel development that replaced storyboards. Progressive designers typically created prototypes with one of two then-new software products: Dan Bricklin's Demo II (for personal computers) or Hypercard® (for the Apple® Macintosh™). These tools enabled designers to draw a monochrome likeness of a device's controls and displays on a computer screen and then add dynamic features to simulate user interactions with the product. For example, a prototype's switches could be programmed to move or its display content could be changed in response to keyboard or mouse inputs. Such prototyping tools offered designers their first opportunity to explore alternative user-interface designs relatively easily and at low cost. In fact, the time required to build a computer-based prototype (versus a working model) was reduced to the point that the expression *rapid prototyping* came into common use.

Today, most manufacturers embrace prototyping as an essential part of developing user interfaces. Popular prototyping tools such as SuperCard™, Director, Toolbook, Visual Basic™, and Altia® Design offer high-resolution color graphics plus advanced image and data manipulation. These powerful tools enable developers to produce prototypes that look and function like real products.

In an increasing number of cases, however, prototyping is no longer rapid—prototype fidelity has become an end in itself, disconnected from the goal of design refinement. This can increase costs unnecessarily, particularly when a lower-fidelity prototype could resolve design issues. Accordingly, developers should first define their prototyping goals, then develop prototypes to match.

Defining the Purpose

A user-interface prototype can serve one or more purposes that pose varying fidelity requirements.

Exploring Design Alternatives

Instead of simply sketching or describing a user interface, developers can create alternative computer-based prototypes as a way to identify the most promising solutions to design problems. Usually, such prototypes consist of simplified illustrations of the product and its important features. For instance, pop-up menus on the device evoked by a simulated key press can present a list of options but not allow the user to go any further with the menu selection task.

Selling a Design Concept

A limited demonstration of how a product works is a powerful tool for selling a design. A baseline image of the product that includes three to five dynamic features might be sufficient to communicate a user-interface concept to design team members (engineers, programmers, writers, and managers).

Obtaining User Feedback

Prototypes enable developers to show users how a product works, rather than requiring them to imagine the process. Designers can then obtain feedback from one-to-one interviews or from focus groups of 8 to 12 people. Again, a baseline image with three to five dynamic features might provide sufficient fidelity.

Assessing Product Usability

Prototypes are a boon to usability testing. Previously, usability testing could not be conducted until the first physical prototype was available, which left little time for design changes based on test results. Instead, changes were often limited to minor, superficial enhancements, which is one reason so many products with obvious user-interface shortcomings have been marketed over the years. Now, computer-based prototypes with enough features to enable users to perform realistic tasks have fundamentally changed the development process by permitting earlier and more frequent testing.

Specifying a Final Design

Many developers are renovating their methods of design specification. Traditionally, a design specification has been a lengthy, wordy document that frequently made it difficult for anyone but its authors to picture accurately how the final device might look and function. As a result, many devices have not lived up to the original vision of their designers, largely because their specifications left many aspects of the user interface up to the implementer's interpretation. User-interface prototypes solve this problem by delivering to engineers and software developers a dynamic model of a device that leaves very little room for misinterpretation.

Once developers have defined prototyping goals, they can begin to deal with fidelity issues, such as whether the prototype should look exactly like the final device (i.e., rendered in full color and three dimensions) or be simpler (e.g., line art); whether users should be able to perform all possible device tasks or just a representative sample; and whether the prototype's response time needs to be perfectly matched to that of the finished device.

Appearance

Although early prototyping tools were graphically limited (i.e., they lacked color, provided few fonts, and presented screen-size limitations), the latest crop of tools are quite powerful. Objects can be drawn as line art (black and white only) or in halftone (grays are simulated by patterns of dots), gray scale, or full color. Product images can approach photo-realism. In fact, a scanner can be used to import high-fidelity images, such as renderings produced by industrial designers.

In a recent test (Wiklund et al. 1992, 399–403) researchers created four prototypes of an actual electronic dictionary, with varying degrees of image fidelity. The product incorporated a QWERTY keyboard and an LCD display, as do many medical devices. Figure 22.1 illustrates two of these electronic prototypes. A sample of 50 users performed several tasks, such as checking the spelling of a word and looking up synonyms, with one of the prototypes and with the real product. Test subjects then rated the prototypes and products according to several criteria, including ease of use.

Test data showed no significant differences among the ratings of the four prototypes based on perceived ease of use, although there were significant differences in terms of preferred appearance. As one would expect, subjects rated products progressively higher in terms of aesthetic appeal as image fidelity increased from line art to a three-dimensional, color rendering. This finding suggests that prototype appearance is less important than developers might think with respect to evaluating usability. Less-refined images are adequate unless color and three dimensionality provide important sensory cues or communicate additional information by virtue of some coding scheme (e.g., red for emergency). If the extra time required to produce a high-fidelity image is trivial, however, the issue is moot. Of course, if part of the purpose of prototyping is to assess users' aesthetic preferences, designers should aim for maximum image fidelity.

Functionality

A user-interface prototype must enable users to perform some sort of task; otherwise it is just an electronic rendering. Deciding exactly how many interactive features to build into a given prototype, however, challenges novice and experienced

Figure 22.1. *Elementary (top) and more refined (bottom) versions of electronic prototypes of an electronic dictionary.*

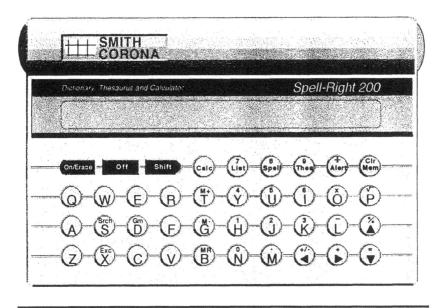

prototypers alike. Again, the goals of prototyping must be closely considered.

In general, prototypes developed early in the design process must be sufficiently functional to enable the developer to evaluate their merits reliably and relatively quickly (e.g., in a usability test). Prototypes of detailed designs usually require more development time and are more functional than prototyped concepts, but still may not be complete models. At both stages it is important to reveal a device's essential features without giving in to the temptation to prototype the entire product.

Jacob Nielsen has developed recently a useful vocabulary for discussing the interactiveness and completeness of prototypes. He distinguishes between horizontal prototypes, which give an overview of a device's essential elements, but do not enable users to perform realistic tasks; and vertical prototypes, which focus on a limited number of features, but enable users to perform sequences of tasks.

> Horizontal prototypes reduce the level of functionality and result in a user-interface surface layer [showing many features], while vertical prototypes reduce the number of features and implement the full functionality of those chosen (i.e., we get a part of the system to play with) (Nielsen 1989, 395).

He also uses the term *scenarios* for prototypes that present only a few of a device's features, with limited functionality. Scenarios require users to follow specific, previously planned paths, limiting both their ability to explore and the developer's ability to assess interaction errors (i.e., false paths). A benefit of a scenario is that because it is usually a limited prototype, it can be changed quickly.

Because they are developed in the context of real-world demands and trade-offs, however, actual prototypes rarely fit into a single class. At the concept development stage, for example, it is unusual to find a purely horizontal or vertical prototype because designs are rarely developed in sufficient detail. Scenario prototypes are more common at that stage but are not substantial enough to support a reasonable usability test; they are best reserved for in-house design explorations.

Ultimately, usability tests require hybrid prototypes that reflect the current level of design understanding and refinement, interesting or contentious elements of the user interface, the skills of the people building the prototype, and the project time available. A practical method for developing a hybrid prototype is to begin with a horizontal prototype that provides a good sense of the product as a whole. This approach benefits medical product users, who seem to be most comfortable dealing with a complete product (if only at the surface level) rather than having to imagine it. If, for example, the prototype device were a patient monitor, the developer would first produce the top-level display, containing vital-sign waveforms and numbers, plus top-level controls, such as those that evoke on-screen menus. At the concept development stage developers might add a few vertical components that simulate frequent, urgent, critical, or complex tasks, or focus on problematic user-interface components. For example, if patient demographics are to be entered into on-screen forms using a virtual keyboard (a computer-generated image of a keyboard) and the process might prove unbearably slow, a developer can prototype this task to check its feasibility.

The vertical components of a patient monitor might include the following:

1. A dynamic electrocardiogram setup menu that enables users to change the waveform height, waveform speed, and lead.

2. A cardiac output worksheet that enables users to initiate a cardiac output measurement, perform specific calculations, and compare data.

3. An alarm adjustment overlay screen that permits users to change the high and low alarm limits for the heart rate, CO_2, and O_2 parameters.

4. An oxygen saturation trend display with a moving pointer that enables users to find a parameter value that appeared 20 minutes before.

Some may feel that this example offers too much detail for a concept-level prototype. However, the quality of user feedback on designs is closely related to the extent of user interactions with devices. If the vertical components are chosen carefully, the effort required to simulate the associated tasks with a moderately high level of fidelity is worthwhile, compared to prototyping a higher number of vertical components with less fidelity. Also, the increased fidelity avoids the pitfall of presenting a design too simplistically, thereby leading to test results that underestimate the usability problems of a given user interface.

The detailed design stage calls for more-complete proto-types, which enable users to explore interfaces more freely and provide developers with substantial insights into how new users interact with their products and where design flaws may exist. Detailed design-level prototypes incorporate many more vertical components than do concept-level prototypes.

Responsiveness

Frequently, a prototype responds more slowly to user input than the final product does. This can create problems if developers want to simulate time-sensitive tasks or if the prototype's slug-gishness substantially disrupts the way users naturally perform a task. For example, several new medical devices, such as pulse oximeters, incorporate a control wheel that moves a cursor among several menu items displayed on a screen. The advan-tage of this feature is its speed—on the real device, when the user turns the control wheel, the cursor moves rapidly among the selectable items. However, several popular prototyping soft-ware tools bog down when simulating this feature. Users click with a mouse on the right side of a virtual control (simulating a clockwise turn), and the display may take a half-second or longer to respond, which is disruptive to performing the task in a natural manner. It is not clear, however, how this disruption might influence overall preferences among alternative input devices.

More research is needed, then, to determine how respon-siveness influences test subjects' perceptions of usability. While common sense suggests that respondents would find a minor response-time lag more tolerable when evaluating concept-level prototypes (as compared with those at the detailed design level), such inaccuracies might conceivably kill a promising design concept.

The best advice is for designers to pay attention to device response time in all prototype development. Some prototypes might need to be slowed down to match the performance of the actual product. This is not usually a problem, because timers or delays can be added to the prototype's underlying code. Speeding up a prototype, however, can be more difficult. Possible solutions include using a faster computer, restructuring the software code, temporarily disabling certain prototype func-tions, or modeling the user-device interactions in a more basic manner. When there is no practical way to match the response time of the prototype to that of the final product, developers should take care to clarify the differences to people participating in evaluations.

Manufacturer Experience

According to Eric Smith, a user-interface specialist with Datex Medical Instrumentation (Tewksbury, MA), prototyping played an important role in the development of the company's AS/3 anesthesia monitor, introduced to the market in 1992. Figure 22.2 shows an early prototype of the AS/3 monitor and the final version. Important changes in the final design included an increase in the menu height to allow for more choices with maximum readability; the elimination of demarcation boxes around numbers to reduce visual clutter; and the use of larger numbers to increase the conspicuity and readability of important information. Note that the prototype and final version are identical in terms of visual exactness, which improves the quality of user feedback on the aesthetics and readability of early design concepts.

> Prototyping provided us with a fast way to explore user-interface concepts and communicate them to other members of the design team. In the past, we relied on black-and-white storyboards to illustrate elements of the user interface. This approach was not well suited to describing the dynamic characteristics of the AS/3 monitor's color displays or to getting user feedback.

Today, Datex views prototyping as an integral step in the user-interface development process.

Smith feels strongly that image fidelity is important to users of Datex products—mostly anesthesiologists and nurses.

> Users need to feel they are looking at a real product. A prototype with a more-primitive appearance does not have full credibility in their eyes and could reduce the quality of their design feedback. By comparison, a lower-fidelity image is fine for an in-house design review, because engineers, marketers, and sales managers can easily tell how the final product might look.

Apart from issues of visual fidelity, Smith says he limits the functionality of concept-level prototypes in order to keep the

Figure 22.2. *An early prototype of Datex Medical Instrumentation's AS/3 anesthesia patient monitor screen (top), which illustrates an invasive blood pressure setup menu. The final version of the screen (bottom) overlays waveform and numeric information.*

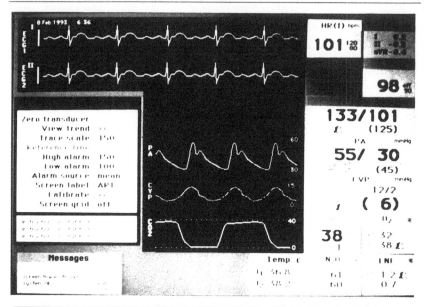

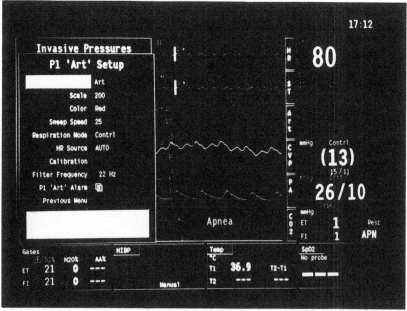

Courtesy of Datex Medical Instrumentation, Tewksbury, MA.

focus on concept-level issues, rather than bogging users down in design details.

> When acquiring user feedback—for example, during a usability test—it is important that concepts share the same level of functionality. Otherwise, you introduce bias to the test.

Smith attempts to avoid discrepancies in response times between prototypes and actual devices when such discrepancies might interfere with usability evaluations.

> I'm not so concerned with menus that pop up or scroll slowly because of the characteristics of the prototyping tool. I know menu interaction speed will not be a problem in the final product, and the prototype still lets me effectively assess the menu designs. I get more concerned about discrepancies in response time with our pointing device [a control wheel], since so many tasks are performed via the wheel and a slow response to inputs [clicks on a virtual control wheel] would frustrate users and influence their opinion of the product.

> Putting a lot of time into building high-fidelity prototypes is less of a problem if you can use some of the software code later to produce working versions of the product.

Some applications software, such as Visual Basic™ (which supports Windows-based application development), are well suited to prototyping needs and also to producing useful code. In the future more prototyping tools undoubtedly will have this capability—a means to further accelerate the design process and reduce waste at the same time.

Conclusion

Prototyping tools are to the user-interface designer as clay is to the automobile designer: an important medium for expressing design concepts rapidly, as well as presenting and validating more-advanced designs. Prototypes also give users the opportunity to evaluate and comment on product designs well before they are finalized, which improves the chances of detecting usability problems before products are brought to market.

The issue is no longer whether to prototype, but how to prototype. Clearly, no user-interface prototype has to resemble the final product exactly in terms of its appearance, function, and response time. Developers need to examine carefully how they use prototypes to ensure that they expend the right level of effort—after all, creating a computer-based prototype should not require the effort needed to build a working model. Furthermore, prototype fidelity should not be regarded as a measure of a prototyper's proficiency. Often, it is perfectly fine for a prototype to be a simple line drawing, simulate just a few tasks, and respond slowly. Matching the fidelity of a user-interface prototype to its purpose will save time and keep the design focus where it belongs—on improving the user interface.

References

Nielsen, J. 1989. Usability engineering at a discount. In *Designing and using human-computer interfaces and knowledge-based systems*, edited by G. Salvendy and M. Smith. New York: Elsevier.

Wiklund, M., C. Thurrott, and J. Dumas. 1992. Does the fidelity of software prototypes affect the perception of usability? In *Proceedings of the Human Factors Society 36th Annual Meeting*. Santa Monica, CA: Human Factors Society.

Testing the
User Interface

Chapter

23

Usability Tests of Medical Products as a Prelude to the Clinical Trial

Most medical product developers know the purpose of a clinical trial and the process for conducting one. Only a small fraction, however, are equally familiar with usability testing. Both product assessment techniques determine whether a product is ready for market and whether it will effectively and reliably serve users' needs. However, marked differences exist between the two techniques in their usefulness for determining whether a product suits its users well. The clinical trial and usability test act as filters, isolating product usability problems (Figure 23.1). Both filters are needed to cleanse the product. Unfortunately, most product developers conduct no usability tests. Lacking this critical filter for usability problems, many products come to market with user-interface design flaws that escaped detection during the clinical trial and penalize end users.

Figure 23.1. *The process of cleansing a product of usability problems.*

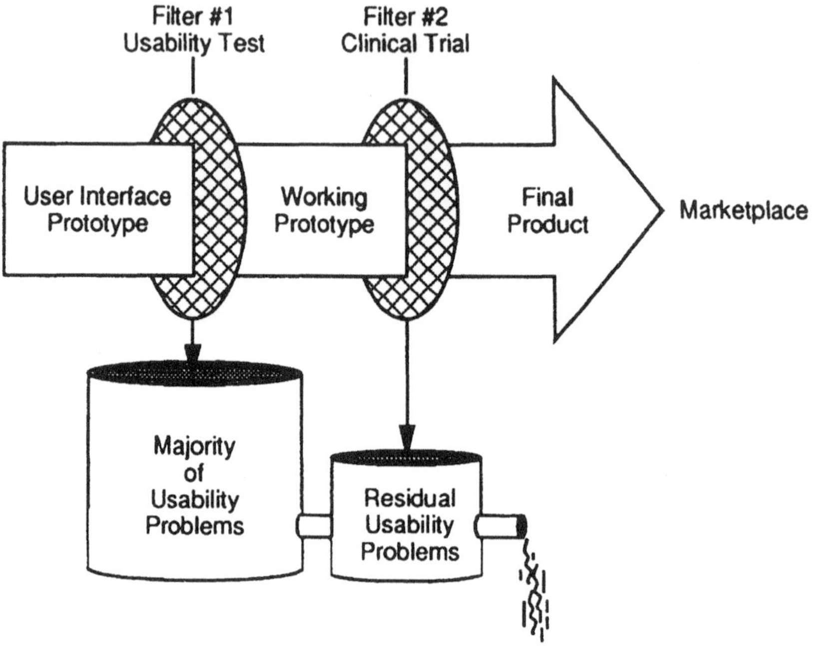

The Clinical Trial

Typically, a manufacturer conducts a clinical trial to ensure that its product performs well in the intended-use environment. Thomas Horth, a manager in Hewlett-Packard's Medical Group (Andover, MA), says,

> The clinical trial serves the primary purposes of demonstrating the medical efficacy of the product and meeting regulatory requirements. Of course you keep your eyes out for usability problems, but catching them is really an incidental benefit. Even if you catch a problem, you probably can't do a lot to fix it, considering that you need a practically finished product before you can conduct a clinical trial. This is why we perform extensive product testing before we go to clinical trial.

Additional objectives of a clinical trial may include collecting information enabling the manufacturer to fine-tune its marketing strategy and, for a computer-based product, finding software bugs.

A clinical trial may be performed at one or more hospitals, clinics, or associated medical practice settings. Most often, it is managed by R&D personnel, drawing support from marketing to handle logistics. However, some manufacturers may take the opposite approach, putting marketing in control of the process. At the start of a clinical trial, a new product may be introduced to users working at the host organization via in-service, where the manufacturer's representative presents the product and walks through its operating procedures. Then, for a planned period the newly introduced product may be used in parallel with the one it will replace. When the users and manufacturer develop enough confidence in the new product, it may advance to full-time use. Of course, the product may immediately be used full time if it is the first of its kind or if no similar products are in use at the clinical trial site.

Over the course of the clinical trial, a developer may collect several types of product performance information and solicit comments and opinions from people who have used the product. This information may be collected through questionnaires, interviews, or a focus group. Often, information about the product's usability is collected less systematically, as in casual conversations between a manufacturer's representative and a user. The trial may span several months or, in special cases, years. From a usability standpoint, clinical trial results provide a valuable but potentially incomplete picture of product performance. Despite the shortcomings of using clinical trials to gather product usability data, determining how users interact with the product in the real-use environment is essential to validating the user interface.

The Usability Test

A usability test differs significantly from the clinical trial, but is equally essential for producing a successful, usable product. This test spotlights the user interface and users' reactions to it. A usability test may take up to a week, depending on the number of users involved. Such tests can be conducted in an office-like setting, away from the medical practice environment. This eliminates interferences that would occur in the actual-use environment. While usability test formats vary, typically one individual at a time performs self-exploration as well as directed tasks with a product. Test administrators can provide special prompts and feedback as required to add realism. As the subject performs tasks with the product, researchers observe and record results. The process gives subjects time to concentrate on using the product. Actually, users may spend weeks learning to use a product. Whether they encounter operating difficulties or causes for dissatisfaction over this time depends largely on how much they use the product and which tasks they perform. A usability test compresses the initial use experience into a shorter time frame, usually one to four hours.

In hunting for usability problems, researchers ask subjects to talk their way through each task, describing what they are thinking, decisions they are contemplating, irritants, advantages, and so on. Sometimes, usability problems surface immediately when a subject tries to turn on a device and cannot find the power switch. In such a case he or she might say:

> Now, I'll turn the power on. I am looking on the front panel but nothing jumps out at me. I see a switch labeled "standby," but I don't think that turns it on. You probably press that to save power without turning it off. I'm reaching around the back for a switch, but I don't feel anything. I would expect to find a switch right here [subject points to lower right side of control panel]. This green light probably illuminates when you turn the power on. Oh, I see [subject presses the light]. This light is the switch. You press it in to turn the power on. Boy, that wasn't obvious to me.

The effect of stress on how people use a product can be studied by introducing time limits, removing product instructions or the user manual, and introducing product failures. Researchers can create a worst-case scenario and see how users react. Test outcomes can be compared across several subjects. Companies performing such tests commonly find that researchers collect a large set of usability problems that might have escaped detection during a clinical trial, since such trials do not explicitly address usability.

Talking aloud can disturb the natural interaction between a subject and a product, so an alternative protocol should be available. Subjects can watch themselves using a product on videotape and describe, retrospectively, what they were thinking and doing. Researchers at Virginia Polytechnic and State University compared the concurrent verbal protocol to the retrospective verbal protocol (Bowers and Snyder 1990, 1270–1274). Forty-eight subjects performed 12 tasks each, using a window-style software system operating on a large and small computer monitor. Interestingly, subjects who talked as they worked gave the richest procedural information (descriptions of the steps they took and the rationale). Subjects reviewing their past performance using a product provided better comments on product design.

When to Conduct a Usability Test

A usability test may be conducted at several points during product design. The rigor of the test depends on the product's completeness, the realism of task scenarios, and recruitment of appropriate test subjects. Generally, it is most productive to conduct usability tests on existing products, user-interface prototypes of new product concepts, and working models that approach the look and feel of a ready-for-market product.

Evaluate Existing Products

When a design process begins, manufacturers may elect to test the usability of an existing product. This may be one they plan to replace or a competitor's product. The results of such a benchmarking test give developers a sense for how they can improve the existing product. For example, if plans are to develop a new infant incubator, the manufacturer would benefit from knowing if a majority of users

- Take too much time to prepare the device for use,
- Are confused by the existing product's temperature controls,
- Dislike the way its door opens and closes, and/or
- Prefer better handles for moving it around.

Test results can motivate designers who are out to beat the competition. For example, if the incubator exposed to usability testing required five minutes for setup, designers may search for ways to reduce setup time to three minutes. Such a reduction would satisfy users and give the manufacturer a strong marketing claim.

Evaluate User-Interface Prototypes

Usability tests may be conducted on early design alternatives, assuming designs have sufficient detail and can provide user feedback. User-interface prototypes developed on computers are especially well suited for such testing (see chapters 21 and 22). User-interface prototypes let people interact with controls and give feedback, such as control movement or changes on a display. Developers may submit competing design alternatives and obtain users' reactions to them via a usability test. Subjects may be asked to perform the same set of tasks with each design. Running several subjects and varying the order in which the designs are presented can counteract learning effects that might improve subjects' performance as they work. During such testing, a manufacturer developing a blood-testing device might compare the usability of three alternative solutions that employ push buttons, rotary knobs, and soft keys (keys with associated labels appearing on a computer-driven display).

Wolfgang Scholz, vice president in charge of technology and product with Siemens Medical Electronics (Danvers, MA), says Siemens traditionally relies on feedback from focus groups in which clinical experts discuss the good and bad points of a design. They now simulate or rapid prototype the product's user interface on a PC screen to obtain early feedback.

> We have experts review the design in prototype form and ask for their recommendations. This allows us to solve most user-interface problems before we start coding software. For example, we might learn that the sequence of keystrokes is poor, so we fix the problem.

Figure 23.2 shows the Siemens Sirecust 1481T telemetry station that evolved from clinicians' feedback on prototypes. Scholz plans to have clinical experts interact more extensively with prototypes as a means to identify and resolve usability problems.

Evaluate Working Models

The best time for conducting a single usability test on a product is when the first working model becomes available. The term *working model* implies that the product can perform most of the final product's functions, has a representative appearance, and is subject to further design refinement. The design is not yet frozen. A usability test employing a working model enables one to study both mental and physical user interactions with the product. Imagine a productive test of a new fluid collection canister used primarily by nurses. On the direction of the usability test administrator, nurses could make the required connections

Figure 23.2. *The Siemens Sirecust 1481T telemetry station.*

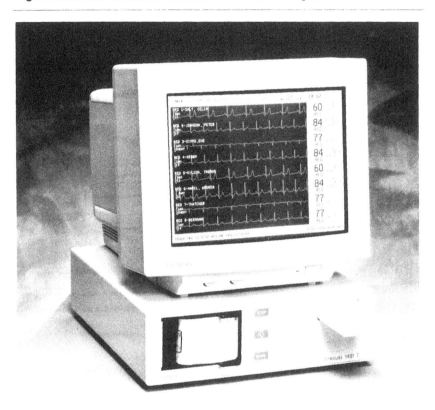

Courtesy of Siemens Medical Systems—Electro-Medical Group.

between the canister and tubing in preparing to suction a patient's lungs. Such a test may demonstrate that using the new canister is intuitive and preferred to existing products, or that subjects become confused when trying to make the right connections. In this hypothetical case a developer might learn that the canister needs better labeling, color or shape coding of ports, and a conspicuous warning.

Early in the design process the manufacturer of a blood-test device might establish quantified usability goals for its product, then test against these once a working model becomes available (see chapter 8). Possible goals for such a blood-test device are as follows:

- Upon initial use 90 percent of subjects shall deposit the drop of blood in the correct position on the test strip.

- After three practice trials 80 percent of subjects shall complete the blood test within three minutes.

- Following instructions, subjects shall require an average of two minutes to replace blood-test strips.

- After a 20-minute trial using the device and its leading competitor, 75 percent of subjects shall prefer the product under test.

Appropriate values would be factored into these goal statements based on studies of existing products and users' needs and preferences.

Validate the Final Product

When product design is complete, a usability test may be repeated to confirm that earlier problems have been resolved. This test may include an integrated or separate evaluation of learning tools, such as a user manual or instructions accompanying the product. A test finding that users experienced no significant problems with the given product is welcome news to technical and marketing people who may have feared a usability problem arising one month after product introduction. If the news is bad (i.e., problems persist), last-minute, superficial design changes, modifications to user documentation (learning tools), or increased training to avert the usability problem might remedy the situation. In addition to reducing paranoia, such a test may also generate marketing claims. While many manufacturers produce advertising copy claiming their products are user-friendly, few can support this claim with usability test results.

Planning and Conducting a Usability Test

Usability testing requires a plan. The plan may describe test objectives and activities, data to be collected, number and kind of test subjects, and analysis and documentation of test results. Test objectives may include the following:

- Collecting users' suggestions for improving design.

- Determining which design users most prefer among two or more alternatives.

- Developing a baseline on user performance, including the time users take to perform a task, their rate of successful completion, and incidence of error.

The test itself may involve self-exploration and directed task sessions, as described earlier. Self-exploration is especially effective for determining a product's intuitiveness. Confidence in one's product grows when a test subject newly introduced to it uses it effectively without guidance from test administrators. Confidence diminishes when a subject flounders in his or her attempt to use the product without direction. To start a self-exploration session with a product such as an infant incubator, a test administrator may advise the test subject:

> Imagine that you are on duty in a neonatal unit. The incubator in front of you has been delivered by the manufacturer for evaluation by the hospital's nursing staff. It is fully functional. Spend the next 10 minutes or so exploring how the device works. Imagine that you will be placing an infant in the incubator 10 minutes from now. Please talk aloud as you explore. Tell us what you are thinking, and what decisions you are making. Tell us when some aspect of the incubator pleases or disturbs you. Ask for assistance only if it is absolutely necessary to continue the exploration.

At the completion of a self-exploration session, one may administer a questionnaire and interview the subject on his or her first impressions of the product. Collecting these impressions before subjects perform directed tasks is important, since initial impressions may shift with a product's continued use. The questionnaire may ask subjects to rate the product on several attributes of the user interface. The interview may include several open-ended questions, such as the following:

- What do you think of the product so far?

- Was it clear how to operate the device?

- What aspects of the design made learning to use it easy/hard?

- Would you now be confident enough to use the product in a real situation?

The directed-task session usually lasts longer than the self-exploration. Directed tasks may involve one step and last a matter of seconds or they may be more involved, lasting several minutes. The key is to define a set of realistic tasks that will put the design through its paces. The task list should probably include all tasks frequently performed, those with a critical purpose, and those developers expect will give users trouble. Subjects should continue to talk aloud while performing directed tasks; test administrators should record comments, task times, errors, and other performance measures. A computer-based data logging system serves this purpose nicely (Philips and Dumas 1990, 295–299). At the close of the directed-task session, a questionnaire may again be administered, followed by an interview. Depending on the directed tasks, the questionnaire and interview might be administered after each task.

Product developers should observe some, if not all, usability tests. Seeing subjects struggling to use a product is humbling but also motivating. By watching the struggle, developers can immediately generate ideas on remedying observed problems. If the developers cannot attend a test, a videotape will enable them to see how things went. Management and marketing personnel are another potential audience for the videotape (or a highlight tape condensing pertinent test events).

Choosing subjects can be tricky when the user population is diverse. Fortunately, most medical products are used by specialists working in specialized environments. This makes setting subject selection criteria and identifying places where subjects can be recruited relatively easy. Recruiting subjects who have not participated in earlier evaluations of the product or similar products and who have no vested interest in the company developing the product is best. This means clinical specialists working for manufacturers are inappropriate. However, a clinical specialist may be perfect to coordinate finding other subjects.

Ten subjects is a good working number for most usability tests, assuming the user population is homogeneous (Virzi 1990, 291–294). Tests involving 10 subjects should identify 90 percent of usability problems and virtually all major problems. Testing more subjects produces sharply diminishing returns on the investment of time and money. However, if several distinct user populations exist, one may want to recruit 10 subjects for each. Subjects are normally compensated for participating in the usability test with an honorarium of $50 to over $200 for a two-hour test. Well-paid professionals, however, are more motivated by the opportunity to influence the design of a product they may eventually use than by the money. Before subjects are informed about the product being evaluated, they should complete any required confidentiality forms.

Where to Conduct a Usability Test

A productive usability test may be conducted in a simple setting, such as a spare office, or a more extensive laboratory setting (see chapter 24). An office may be equipped with a camcorder for documenting the test. Test administrators may sit in the room with the subject during testing and record test results on a note pad. Although this back-to-basics approach may reduce the quality of data collection slightly, most companies can use it immediately and without making a large investment.

Having a special setting for usability testing has clear advantages, however. A laboratory built for usability testing typically includes two adjacent rooms connected with a one-way mirror. Subjects interact with products in the room that contains recording equipment. Depending on the test administrators may remain with the subjects or communicate with them from the observation room via intercom. Figure 23.3 shows a typical laboratory layout.

Siemens Medical Electronics has even simulated an intensive-care setting (Figure 23.4), enabling it to evaluate the usability of its patient monitoring devices and their compatibility with other equipment normally found in an ICU. For example, Siemens can study the persistent problem of how users cope with all the tubes and wiring connecting patients to attendant devices. The company invites clinical experts to spend time interacting with a monitor and subsequently provide feedback on design. Scholz considers this process a boost to Siemens' ability to bring more usable products to market.

Reporting Usability Test Results

When usability testing is completed, there can be plenty of data to reduce and analyze. Collecting information with structured forms simplifies the job. Questionnaire data such as ratings are well suited to testing. This room may be empty except for the furniture necessary to conduct the test and the cameras

Figure 23.3. *Usability test laboratory floor plan.*

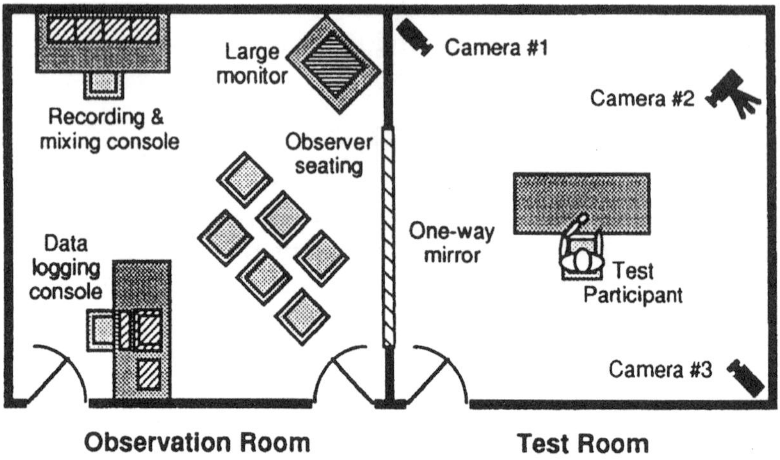

and microphone(s) needed to capture the action. An adjacent room allows several people to observe the test without disrupting it and houses the data-statistical analysis. A simple calculation of means and standard deviations can reveal where a product may need improvement. More rigorous statistical techniques may be applied to compare results across subject types or test conditions. Data of this kind can be presented in chart or tabular form, providing a quantified view of the product's usability.

A running log of the subject's comments during the test can also be revealing. Normally, subjects become outspoken after taking a few minutes to familiarize themselves with the test surroundings. A compilation of their comments often points directly to a product's major weaknesses and may include the subjects' recommendations for correcting problems. When 8 out of 10 subjects complain about the same control placement, for example, the need for a design change becomes irrefutable.

All quantitative and anecdotal information should be consolidated in a usability test report, which should include a summary of test results and recommendations for design improvement. Test methodology and supporting data can be presented in the body of the report.

Figure 23.4. *Siemens Medical Electronics' simulated intensive-care unit.*

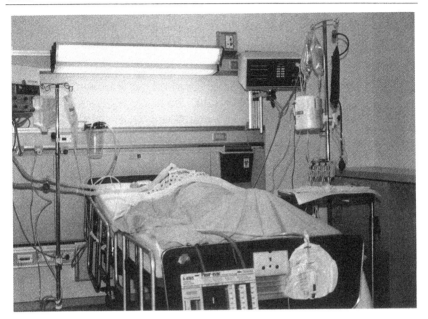

Courtesy of Siemens Medical Systems—Electro-Medical Group.

Implementing the Results

Usability test reports should be distributed to all members of the design team as well as to marketing personnel who will control introduction of the product into the marketplace. Problems cited in the report should be resolved. The pros and cons of fixing each usability problem should be carefully considered. Not fixing a problem may be perfectly reasonable after considering the benefits and costs. The important thing is to follow a rational decision process. In some cases, retesting the revised product may be worthwhile. This is especially true if design changes significantly alter how users will interact with the product. If design changes are minor, user-interface design specialists involved in the design effort may simply give them their blessing. The next step is production of a final prototype that will go to clinical trial, already free of most usability problems.

References

Bowers, V., and H. Snyder. 1990. Concurrent versus retrospective protocol for comparing window usability. In *Proceedings of the Human Factors Society 34th Annual Meeting*. Santa Monica, CA: Human Factors Society.

Philips, B., and J. Dumas. 1990. Usability testing: Identifying functional requirements for data logging software. In *Proceedings of the Human Factors Society 34th Annual Meeting*. Santa Monica, CA: Human Factors Society.

Virzi, R. A. 1990. Streamlining the design process: Running fewer subjects. In *Proceedings of the Human Factors Society 34th Annual Meeting*. Santa Monica, CA: Human Factors Society.

Chapter

24

Building a Usability
Test Laboratory

Traditionally, manufacturers have rigorously tested devices to determine whether they meet corporate performance specifications or standards set by organizations such as the American Society for Testing Materials or Underwriters Laboratories. To determine if it will stand up to conditions of actual use, a device may be alternately chilled and heated, vibrated for many hours, or dropped at various angles from a set height.

Today, some manufacturers extend performance testing to include usability testing—assessing whether device users can perform tasks within established boundaries, such as a time limit. Unlike performance testing, which is conducted just prior to a product's release, usability testing can be conducted at several stages of the design process, such as during concept development. Frequently, usability testing is carried out in any convenient setting, such as a hotel guest suite at the site of a medical convention. Manufacturers that expect to perform extensive usability testing, however, would be well served by building their own usability test lab.

Building a usability test laboratory strengthens an organization's commitment to product usability by enabling manufacturers to involve actual users in the device design process, using either designs in progress (rapid prototypes, working models, or alpha-release products) or production models. Typically, a usability test laboratory includes one room for test subjects and another to observe, direct, and document test proceedings.

Compared with other types of labs, a suitable usability test lab need not be expensive to build or maintain. In fact, it may not even be a permanent facility—two empty offices, a camcorder, and a TV monitor will suffice. However, a moderate investment in a dedicated facility is likely to improve the testing process by enabling those conducting tests to more effectively control and monitor proceedings. A dedicated facility also contributes a sense of legitimacy to the work of the usability testing staff.

Lab Activity

The principal people involved in a usability test are the subject and the administrator; however, additional staff often document or observe the proceedings. Thorough documentation may require someone to log test data and operate video recording equipment. Observers might include R&D managers, market researchers, user-interface designers, software managers, and product documentation writers.

At the start of a typical usability test, the administrator greets the subject, brings him or her into the test room, and explains the planned activities. The balance of the test can progress in various ways. Often, the test subject first spends time exploring the device's user interface with little, if any, direction from the test administrator. Next, the administrator may ask the subject to perform one or more specific tasks with the device. During this test phase the administrator usually sits in the observation room, unless repeated adjustments to the device are necessary. The administrator returns to the test room near the close of the session to conduct a posttest interview. A more complete discussion of the usability testing process is presented in *A Practical Guide to Usability Testing* (Dumas and Redish, Ablex Publishing Corp., 1994). Illustrations and floorplans of several usability testing laboratories can be found in *Usability in Practice: How Companies Develop User-Friendly Products* (Wiklund, Academic Press, 1994).

Lab Layout

Many usability test laboratories are created from preexisting workspaces, such as two adjacent offices. Figure 24.1 shows a layout consisting of two 13 × 16 foot rooms connected by a one-way mirror. The test room can be configured to simulate a wide range of working environments, such as a doctor's examining room, a blood testing lab, or a portion of an operating theater. Such setups can become quite elaborate when conducting an accurate usability test requires a realistic user environment—for example, to evaluate how users manage the large number of patient-monitoring cables and tubes in intensive-care environments.

In the observation room in Figure 24.1, two separate workstations needed to document testing procedures have been positioned so that staff can intermittently watch the video monitors and observe the test subject through the one-way mirror. Both workstations point away from the one-way mirror so that the computer and monitor displays do not produce distracting reflections or become visible from the test room side of the mirror because of their high luminance. The room layout shown provides space for four observers.

Figure 24.1. *A basic usability test lab constructed from two adjacent offices.*

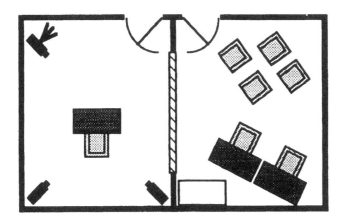

Figure 24.2 shows an alternative, two-room layout that provides a segregated seating area on a raised platform to accommodate a larger number of observers. Another interesting
feature of this layout is the window into the observer area (right
diagonal wall on the figure) so that the testing activity can be
showcased.

Figure 24.2. *A two-room usability test lab with a segregated, raised
seating area. Note window wall to enable visitors to view testing
activity.*

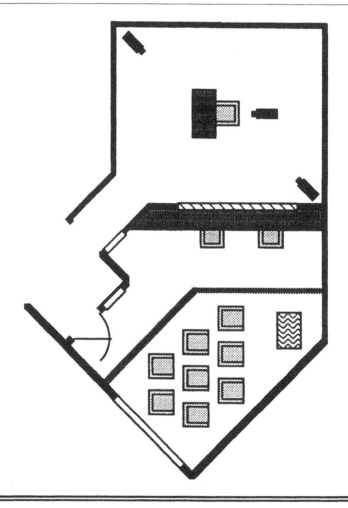

Accommodating an even larger number of observers would require either a larger observation room or committing a third room to observation. Figure 24.3 illustrates a more elaborate setup, which provides observers their own workspace. Note that the second observation room does not share a wall with the test room. This configuration is particularly beneficial in situations where administrators wish to minimize the transmission of sound between the test and observation rooms. Another advantage to this setup is that it enables two tests to run in parallel.

A drawback of this configuration, however, is that it reduces the observers' sense of immediacy, since direct viewing through the one-way mirror is replaced by viewing one or more video images. Another drawback is that test personnel and observers are isolated from each other and cannot easily confer

Figure 24.3. *An elaborate usability test lab setup, with a third room that completely isolates observers from the testing room.*

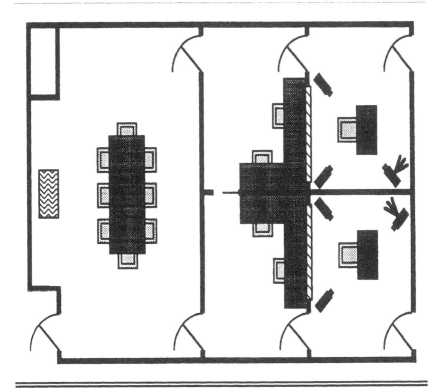

about the test in progress—suggesting questions to ask the subject and answering technical questions that may arise.

Ideally, a usability test laboratory should be located close to the building's entrance, perhaps near a waiting area, and in proximity to a restroom. It should not be located within R&D or manufacturing workspace, since walking through such workspaces might intimidate test subjects or inadvertently reveal confidential information to them.

Equipment and Construction

A complete list of the features and equipment that might be found in usability test laboratories is extensive, but it is not necessary to include them all in order to produce a useful test setting. The level of sophistication required of a lab depends on corporate resources and the extent of testing activities planned. Lab construction will likely require the services of carpenters and electricians, as well as specially trained audio and video technicians. Companies that sell professional-grade audiovisual equipment frequently provide technicians who can install and maintain it, and train test personnel to use it. Consider including the following features in a usability test lab.

One-Way Mirror

A one-way mirror enables observers to view test proceedings without distracting the subject. A mirror size of 10 ft wide by 4 ft high provides a satisfactory view. The most economical and practical way to provide this large a viewing area is to mount two mirrors, each 5 ft wide by 4 ft high, side by side, as shown in Figure 24.4. A one-way mirror can be purchased from most glass suppliers at a cost of about $300 per 4 × 5-ft sheet. Typically, the mirror is framed with wood molding on all sides. Resilient padding should be placed between the molding and the mirror surface to protect it from stress that could crack the glass.

A low-cost alternative to installing a one-way mirror is to observe tests solely by means of a large color video monitor linked to a video camera in the test room. However, this approach creates the aforementioned problem of diminishing the connection between subject and observer and of limiting the observers' view to what the camera sees.

It may at times be necessary to provide test room privacy, especially if the room is used for other purposes, such as conducting focus groups, user interviews, or design critiques.

Figure 24.4. *A view of the testing procedure through the one-way mirror in the observation room at the American Institutes for Research usability test lab.*

Privacy can be created easily by installing a set of blinds over the one-way mirror.

Soundproofing

It is frequently necessary for administrators and observers to converse during a test, which could disturb test subjects. To prevent this, the test room should be acoustically isolated from the observation area. Adding a layer of quarter inch safety glass with embedded plastic to the observer side of the mirror will reduce sound transmission. The cost is about $150 for two 4 × 5-ft sheets. Before sandwiching the mirror and glass, meticulously clean both inner surfaces.

Another way to increase soundproofing is to add a layer of sound-deadening board or sound insulation batts to the partition walls around the test room. If the room has a suspended ceiling, the sound-deadening board or batting should be extended to the superstructure above (i.e., to the concrete or metal decking). A high-grade acoustical tile ceiling will further reduce sound transmission.

Sound System

An assortment of microphones may be required to enable a full range of testing activities. A ceiling-mounted microphone is preferable when subjects will be required to move around the room; a clip-on or freestanding table microphone is best if subjects will remain in a seated location.

The observation room can be equipped with a sound mixer that enables test personnel to create a voice-over on the videotape of test proceedings. This is particularly useful for creating a highlight tape of a usability test.

An intercom is needed for communications between the test room and the observation room (sound-mixing panels sometimes come equipped with an intercom). To avoid acoustic feedback, the intercom speaker should be placed at an appropriate distance and pointed away from the microphones. The intercom speaker's sound level should be adjustable from the observation room. Ideally, the sound system should be wired so that comments made over the intercom are captured on tape and that the test participant and the test administrator can conduct two-way conversations comparable to conversing by telephone.

A telephone line should be installed in both the test room and the observation room. This facilitates calling into the test room to order a lab test, having a test subject call a fictitious hotline, or providing a modem connection. The phone can be equipped to flash a light, instead of ringing, so that it is not overly distracting during a test.

Lighting

For a one-way mirror to function properly, the lighting level in the observation room must be substantially lower than in the test room. An inexpensive solution is to place incandescent lights on a dimmer switch, but a better solution is to install dimmable fluorescent lighting. Fluorescent lighting fixtures should be equipped with glare-reducing louvers, particularly if testing products with computer displays. It may also be useful to install dimmer-controlled track lighting to provide spot lighting of work surfaces.

Climate Control and Ventilation

The observation and testing rooms can become overheated from test personnel, observers, and participants, and from any audiovisual or other electronic equipment in the room. Adjustable air conditioning vents in the ceiling, a recirculating fan, or placing the rooms on independent thermostats can help to accommodate this situation.

Cables and Cable Conduit

A large assortment of cables are normally required to make a suite of audio and video equipment operate properly. To reduce electrical noise in the associated signals, all audio and video cables (especially microphone cables) should be shielded.

Since requirements change, it is impossible to anticipate all the types of room-to-room cabling that may be required to conduct usability tests. Therefore, it is a good idea to install an oversized cable conduit between rooms (with multiple access ports), both to carry audiovisual cables and to allow for future cable needs.

Computer Network Connection

If a device must be connected to a computer network, the appropriate connectors should be installed in the laboratory rooms.

Electrical Outlets

In a usability testing lab, innumerable devices have to be plugged in. For maximum flexibility, strip outlets can be mounted near the floor level along the test room and observation room walls. If tests involve equipment sensitive to power fluctuations, the outlets should be connected to a power conditioner.

Video Equipment

A conventional camcorder mounted on a tripod may be adequate for documenting a simple usability test, particularly if the item of interest remains in the same location. However, a better setup is to install three professional-grade, remotely controlled video cameras, permanently mounting two on the wall or ceiling and the third on a tripod for placement flexibility. The permanent mounting positions should be tested before committing to them. Needed videocamera adjustments include tilt, pan, zoom, manual and automatic exposure level, and manual and automatic focus.

If you plan to conduct extensive tests of devices that incorporate a computer screen, you may want to include a scan converter on your purchase list. A scan converter solves the problem of capturing a flicker-free video image of a computer screen. Instead of training a videocamera on a computer display, a scan converter takes the image signal directly from a computer's external port and converts it to a TV-compatible, NTSC signal that can be fed directly into a video monitor or recorded on videotape. Be aware that some lower-cost scan converters may not work with all types of computers.

The observation room should be equipped with a separate monitor and recorder for each video camera, as well as a monitor for displaying images generated by a mixing panel, if one is installed. While small, black-and-white monitors may be adequate for controlling video cameras, larger color monitors give test observers a better view of the test proceedings. A large monitor can be placed on a tall, rolling stand.

In order to both record and display the video camera images, separate video recorders are required for each video camera. It usually suffices, however, just to record the image produced by a video mixer.

A video mixer can receive an image from any of the video cameras and display it on a monitor. A mixer also allows images from two cameras to be combined. For example, a cropped image of the test subject's head and shoulders can be inserted in the upper right corner of a close-up image of the product in use. This capability makes videotapes much more informative and interesting to watch. A video editor is necessary for making highlight tapes of usability test sessions.

Ideally, all video equipment installed in the observation room should be mounted in a console-style workstation.

Data Logger

In simple usability tests logging data can be accomplished by taking notes by hand and timing events with a stopwatch. For more extensive tests, however, a better approach is to use a specially designed data-logging program that runs on a microcomputer, such as TestLog'r™ (Usability Sciences Corporation, Dallas, TX or OCS TOOLS™ (Triangle Research Collaborative, Inc., Research Triangle Park, NC). Such products enable users to establish codes for specific events, such as a user error or request for assistance. This makes it easy for test administrators to record events, along with an automatic time stamp that can synchronize the clock or counter shown on the videotape, in a kind of shorthand. Many companies end up developing data-logging software tailored to their specific needs (Philips 1990, 295–299).

Seating

Considering the long periods test subjects, personnel, and observers can spend in usability testing, high-quality, ergonomic seating is important. The observer seating area can be terraced to provide a large number of observers with a clear view through the one-way mirror. It is also useful to provide a few small tables for use as writing surfaces and for miscellaneous purposes. A built-in closet or stand-alone cabinet can be helpful for storing ancillary audiovisual equipment and other test supplies.

Costs

The cost of constructing a usability test laboratory depends largely on whether existing office space is being converted or new construction is being undertaken, and on how completely the lab is equipped.

Minimal Configuration

A company that owns a camcorder, a video monitor, and a long coaxial cable is outfitted to run a usability test. Otherwise, the required expenditure for this equipment should be less than $2000. Highlight tapes can be produced using outside video editing services; test data can be recorded by hand.

Moderate Configuration

The electronic equipment required to operate a dedicated testing facility with one test room and one observation room can cost from $20,000 to $30,000 (see Table 24.1). Buying used, professional-grade equipment can lower equipment costs substantially. The cost of architectural improvements is typically about $10,000.

Extensive Configuration

An extensive usability test laboratory can cost $100,000 or more; whether such a high cost is justified depends largely on the volume and importance of the usability testing a company conducts. In addition to its testing function, a well-equipped facility can serve an important public relations and marketing role. A

Table 24.1. *Cost estimate for a moderately well-equipped usability testing laboratory.*

Quantity	Item	Unit Cost ($)	Subtotal ($)
3	13-in. monitor	300	900
1	25-in. monitor	550	550
4	S-VHS editing videotape recorder	1150	4600
1	Edit controller	400	400
1	A/V mixer	2050	2050
1	Character generator	500	500
3	Video camera	1300	3900
4	AC adapter	200	800
2	Pan & tilt mechanism	600	1200
3	8× lens	500	1500
2	Camera mount	50	100
1	Camera body (for tripod mount)	350	350
1	AC adapter (tripod-mount camera)	100	100
1	Tripod	350	350
1	Editing console	650	650
	Assorted cabling	1300	1300
	Installation and training	700	700
Total			$19,950

10-minute tour through an attractive laboratory sends a message to customers that usability is important, and that products are being tested on paid subjects rather than on paying customers.

Conclusion

Usability testing facilities have proliferated within the computer industry—for example, IBM, Microsoft, Lotus, Digital Equipment Corp., and Hewlett-Packard operate their own labs. By contrast, only a handful of medical device manufacturers have their own facilities.

Of course, companies new to usability engineering and testing should not begin their transformation by immediately building a lab. It makes more sense first to find people with training in usability engineering and develop a comprehensive program for improving product usability, which should include usability testing. Initial usability testing should probably be performed using a minimal testing configuration, or by using lab facilities operated by consultants. A successful round of testing, producing important design insights, should be sufficient fuel for a proposal to continue a usability engineering program and building a dedicated test facility.

Reference

Philips, B., and J. Dumas. 1990. Usability testing: Identifying functional requirements for data-logging software. In *Proceedings of the Human Factors Society 34th Annual Meeting*. Santa Monica, CA: Human Factors Society.

Developing Learning Tools and Warnings

Chapter

25

Writing Effective User Manuals

What becomes of the user manuals that are supplied with new medical devices? According to manufacturers and users alike, many get thrown away with the plastic foam and shrink wrap—particularly when a large quantity of devices are purchased at once. Some copies may survive the initial purge, to be partly read by a few department members. Usually, these copies wind up crammed in a file cabinet with user manuals for other devices. When someone retrieves one, perhaps to learn about an unfamiliar feature or solve a technical problem, he or she typically uses it as a reference manual.

Recognizing that user manuals are often used in this limited, task-specific way, many manufacturers assign a low value to them and limit the resources for their development. This results in a self-perpetuating cycle: the manuals produced are lower in quality; users find them difficult to understand and become progressively more dissatisfied with them; and, finally, users develop a general disregard for the manuals.

The underlying cause of this cycle may be that medical workers generally prefer hands-on training to printed documentation. Nevertheless, user manuals continue to perform an important function. Some users—particularly part-time staff and workers who rotate among several hospitals—may not be able to attend in-service presentations, leaving them more dependent on manuals. Also, a 20- to 60-minute in-service training session may be inadequate to teach people everything they need to know about a complex product such as an imaging device or blood-gas analyzer. Finally, some people prefer to read a user manual as a supplement or alternative to hands-on training.

A Worthwhile Investment

Despite pressure to reduce costs and come up with alternatives to user manuals (e.g., on-line documentation), some manufacturers continue to invest substantial resources in their development. They are convinced that a good user manual contributes to total product quality and reflects their concern for users' needs. One way these manufacturers improve the quality of their manuals is by inviting end users to participate in the process of establishing the objectives and evaluating the design of their documents.

Paul Nass is supervisor of technical publications at Marquette Electronics in Milwaukee, a company committed to high-quality user manuals. He and his staff of seven writers and one graphic artist develop manuals for anesthesia monitoring devices. Figure 25.1 is an excerpt from the user manual of Marquette's Tram model critical-care monitor. Nass recognizes that users are unlikely to read manuals completely or in page order.

> We don't expect users to sit down and read our manuals like a novel, from front to back. Instead, we organize our manuals so that people can retrieve the most important information quickly.

Marquette's commitment to producing high-quality manuals has paid off in terms of customer satisfaction. In a recent survey of medical services provided by original equipment manufacturers, Marquette's documentation received an average grade of 88, placing it in a tie for first place with Hewlett-Packard (McGillicuddy 1991, 20).

Nass believes strongly that the manual-development process also serves an important quality assurance function.

Figure 25.1. *Excerpt from the user manual of the Tram model critical-care monitor, made by Marquette Electronics. Note that the arrows in the drawing illustrate what is described in the text, that a final sentence (highlighted with capital letters) summarizes the principal instruction, and that the text is written in plain language.*

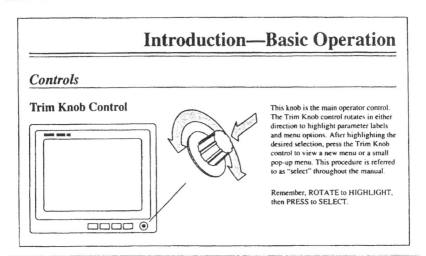

Introduction—Basic Operation

Controls

Trim Knob Control

This knob is the main operator control. The Trim Knob control rotates in either direction to highlight parameter labels and menu options. After highlighting the desired selection, press the Trim Knob control to view a new menu or a small pop-up menu. This procedure is referred to as "select" throughout the manual.

Remember, ROTATE to HIGHLIGHT, then PRESS to SELECT.

Courtesy of Marquette Electronics.

When our writers conduct research on a product under development, they invariably discover usability problems with the hardware and software. Because of Marquette's relaxed organizational structure and management style, the writers are able to take the initiative to resolve these problems, and this contributes to overall product quality and gives them tremendous job satisfaction.

If regulations did not require a user manual, some companies that produce computer-based medical products would prefer to eliminate them altogether. However, experts on paper-based and on-line documentation advise against such a move. Although on-line documentation has great potential to solve certain types of problems by providing easy access to information, "help" keys, and so on, many on-line systems have severe limitations. One limitation is a small-capacity memory. Graphic images require a lot of memory, which might lead system designers to cut down on the number of graphics they use and impair the overall effectiveness of the assistance. Another potential drawback is that users may need information most when on-line help is not available—for instance, when there is a computer failure. Many consumer software companies have resolved the trade-off between on-line and paper-based approaches by providing both. They put quick-reference information on-line, while presenting operational overviews, detailed operating and maintenance procedures, and safety information in a user manual.

Taking a Design Approach

To realize the full potential of a user manual, developers should view it as the end product of a design process rather than a mutated form of a device's functional specifications. One possible model for the creation of a user manual is shown in Figure 25.2. Key elements of this model include analyzing the needs of the target audience as a basis for defining the manual's objectives, applying established document design guidelines to produce a draft of the manual, conducting usability tests of a draft manual on an iterative basis, and gathering feedback on the usefulness of the manual once a medical device is marketed.

The most important step in developing a high-quality user manual is understanding the audience. As Nass says, medical workers rarely sit down with a user manual and read it cover to cover. Doing so would be a boring exercise, and users are busy people who do not normally have the time to read a 100-page manual, which is typical for advanced electronic devices such as anesthesia monitors. They would rather have a sales representative show them how to use the product and keep the manual on hand in case they forget what they were shown.

Moreover, when they do read a manual, medical workers usually seek specific information rather than an overall orientation to a medical device's operation, and user manuals should incorporate this predisposition into their design. Section tabs, an index of key words, illustrations, and a complete table of contents can help users find relevant information quickly.

The user's first challenge is to find the pertinent information. Next, he or she must comprehend and remember it. As a rule, medical workers dislike "technobabble"—jargon-filled narrative that assumes a depth of technical knowledge irrelevant to the

Figure 25.2. *A model for the creation of a user manual. Key elements include analyzing the needs of the target audience in order to define the manual's objectives, producing a draft document based on design guidelines, conducting usability tests, and gathering feedback on the manual after the device is marketed.*

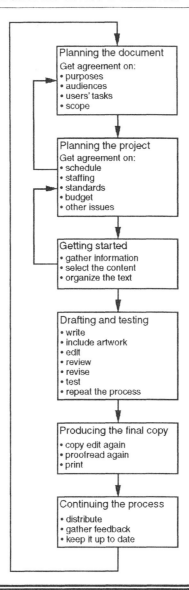

average health-care worker. Such writing makes some users angry with the manufacturer and makes others feel inadequate. To minimize the possibility of such responses, manual writers need to find ways to state technical information clearly and simply.

Some writers say that the manuals they create are frequently complicated and difficult to use because the devices they describe are themselves complex or have poorly designed features. Such claims may be partly true. However, the express purpose of user manuals is to explain things that are not intuitively obvious; writers need to accept the fact that no user interface is perfect. According to Nass,

> A bad user interface may require more work to explain its operation in a reasonable, straightforward manner, but it can be done. It's our job.

To express themselves clearly, manual writers must understand the tasks users need to perform, and present only the information pertinent to those tasks. People formally trained in technical communication are particularly good at this. Task analyses performed by human factors specialists can help give writers a head start (see chapter 7).

Manual writers at IVAC (San Diego), a manufacturer of infusion pumps, find that the best way to develop task-oriented instructions in language easy for users to understand is for writers to work closely with clinical specialists. Susan Cosio, graphics coordinator at IVAC, says,

> Clinical specialists have a model in their heads of how a device is supposed to work. We ask them to write down a procedure from which our writers can develop a first draft of a manual.

In some companies writers use a transcript of the clinical specialists' hands-on instructions to new users as the basis for writing a first draft.

Adhering to Guidelines

Another way to improve the quality of user manuals is to adhere to established document design guidelines. Several textbooks on documentation contain useful guidelines (Doheny-Farina 1988; Horton 1990). Some sample guidelines might include the following:

- Use informative headings.
- Put sentences and paragraphs in a logical sequence.
- Use the active voice.
- Use personal pronouns (e.g., *you* instead of *the user*).
- Avoid using nouns created from verbs; use action verbs.
- Write in short sentences.
- Eliminate unnecessary information.
- List special elements or conditions separately (instead of incorporating them into text).
- Use highlighting techniques, but sparingly.
- Use ragged-right margins (Felker 1981).

Such guidelines may be helpful to inexperienced technical writers. It is also useful to acquire a variety of well-written user manuals as a way to obtain ideas about manual organization, page format, and writing style.

Using Graphics

The visual quality of a document is important for capturing readers' attention and helping them navigate through the contents of the manual. Assuming that the subject matter lends itself to visual presentation, replacing text with graphics reduces the reading load and increases the visual appeal of user manuals. Graphics are especially appropriate for illustrating physical and visual interactions between people and devices, such as how to press switches and scan a displayed image.

Documentation specialist William Horton says:

> In documentation, graphics are no longer an option—they are required. . . . Raised on a diet of television, movies, and videogames, today's user more willingly and skillfully interprets graphics.

According to Horton, the benefits of graphics include the following:

- **Improved job performance.** Graphics that show a logical sequence of actions help users learn procedures more quickly and perform them with greater accuracy. In one study procedural error rates were three times higher for workers who followed textual procedures than for those who learned from graphically depicted procedures.

- **Global communication.** Graphics are often a more reliable method of communicating to an international, multilingual audience; this can be a significant concern for manufacturers marketing products to Europe and Pacific Rim countries.

- **Audience enjoyment.** People who are disinclined to read a lot of text may enjoy reading a document with high-quality graphics.

- **Heightened credibility.** Evidence suggests that people place greater faith in things they see than in things they merely read about.

- **Faster information processing.** Users who read written descriptions often must convert words to mental images before they fully understand an explanation. Graphics can speed their comprehension by eliminating this step.

- **Compactness.** A graphic can communicate more information in less space than words would require. The liberal use of graphics generally results in a manual that is shorter than an all-text version.

- **Compatibility.** Graphics can express concepts such as spatial relationships that either defy description or require a lot of words.

- **Ease of remembering.** People generally remember an image better than they remember words—for example, they normally recognize a person's face more easily than they remember his or her name. The same holds true for remembering an operational sequence diagram compared with an operational sequence description (Horton 1991, 1–11).

Unfortunately, using graphics can be expensive and time-consuming, despite advances in drawing and desktop publishing software that make it relatively easy to integrate text and graphics on a single page. IVAC's Cosio says,

> Our writers face the same pressures to reduce costs as engineers face. We are trying to reduce the cost of user manuals to about $5 apiece; future budgets could be even lower.

Such tight budgets leave little room for professional graphics development, and Cosio's group is fortunate to have four graphics specialists on staff. In cases where a technical publication staff has no graphic specialists, writers—with supplemental training in visual communication—may have to function as artists.

Early commitment is the best way to ensure that the necessary resources will be available to develop professional-quality graphics, as well as to avoid several pitfalls cited by Horton (1991):

- Adding graphics as an afterthought once the text has been written.

- Eliminating graphics as deadlines for camera-ready pages approach.

- Giving less editorial emphasis to graphics quality than to text quality.

Testing the Manual

Traditionally, evaluating manuals involves distributing draft copies to various departments within the development organization (e.g., quality assurance, engineering, marketing, legal, and regulatory affairs). Each department reviews the document to determine whether it meets their acceptance criteria (e.g., consistency with engineering specifications, minimal exposure to liability claims). Reviewers send their comments back to the writers, who integrate them into a revised document. It may be necessary to repeat this process several times before the document is published.

Although this approach to testing user manuals may make use of reviews by clinical specialists, from a human factors standpoint it is seriously flawed because it excludes users from the review process. As a result, the real-world usability of the manual remains unknown.

The solution is to conduct usability tests of draft documentation. Marquette's Nass says,

> We have conducted a limited number of usability tests and plan to do more. In one recent case, we found that users were successful in following the service manual to upgrade the software in an anesthesia monitor. This made us feel comfortable about publishing the manual—with some changes in response to the test findings—and gave our writers much-needed feedback. Usually, you never hear whether customers find the

manuals useful. Testing makes it easier to know whether you're on the right track.

At a minimum usability tests should involve five test subjects, a draft manual, and the associated device prototype or preproduction model (if available). Test subjects should be representative users unaffiliated with the manufacturer. The draft manual should have all major design features in place, although not necessarily in final form—including, for example, a complete table of contents, representative page layouts (including headings), text, and graphics, and a complete index.

The first part of the test typically is devoted to evaluating how well the manual enables users to interact with the device. The administrator may ask subjects to use the manual either on an as-needed basis, reflecting a real-world use pattern, or in a more complete fashion, so that the test will produce more design insights. Initially, the administrator may ask the subject to explore the user interface for about 15 minutes; during this time, the administrator observes in what ways the subject uses the manual.

Then the administrator may ask the subject to perform a series of directed tasks for about an hour, which should include the more important, frequent, difficult, and perhaps obscure tasks required when using the device. Depending on his or her expertise, a subject might refer to the manual a great deal or very little at this time. Tasks might include preparing the device for use, calibrating the device, replacing a reagent solution, or determining the device's cleaning requirements.

The latter portion of the test should focus solely on the subject's understanding of the user manual. The administrator might ask the subject, for example, to find a specific calibration procedure in the manual, determine the default alarm settings, find the section on troubleshooting, or determine the inspection cycle.

Researchers should record the time necessary for subjects to perform tasks, as well as noting errors and any other misunderstandings that arise. A variety of methods and procedures—for example, questionnaires, rating forms, and structured interviews—can be used by researchers to capture subjects' impressions of a manual and to come up with ideas on how it can be improved. Administrators can determine whether subjects remember information in the manual by making sure the material has been read, then administering a comprehension test.

Several document design specialists advise against using read-ability formulas to evaluate the effectiveness of user manuals because such formulas cannot measure whether a manual is sufficiently task-oriented.

Conclusion

Flipping through the pages of a user manual for a few minutes is sufficient to reveal a good deal about its quality. Attractive graphics, informative headings, and balanced text layouts signal that a manual will appeal to users. Of course, a captivating visual style should be reinforced with effective, task-oriented organization of content and a plain-language writing style.

By using high-quality examples as models and following the advice of document design guides, most manufacturers can produce manuals that serve device users well—assuming, of course, that product managers do not subvert the document design process by cutting off the necessary resources. Producing a high-quality manual requires talented writers and graphic artists working as a team, with the support of upper management.

References

Doheny-Farina, S. 1988. *Effective documentation.* Cambridge, MA: MIT Press.

Felker, D. 1981. *Guidelines for document designers.* Washington, DC: American Institutes for Research.

Horton, H. 1990. *Designing and writing on-line documentation.* New York: John Wiley & Sons.

Horton, H. 1991. *Illustrating computer documentation.* New York: John Wiley & Sons.

McGillicuddy, J. 1991. Making the grade. *Second Source Biomedical* November/December:20.

Chapter

26

Creating Quick-Reference Guides for Medical Devices

Because medical workers are busy with direct patient care, they value shortcuts to solving device-related problems, such as finding time to learn to use a new medical device. For example, a physician, nurse, or technician might need several hours to learn the details of how an advanced patient monitor works. The learning procedure might include attending an in-service training session conducted by the manufacturer, reading a lengthy user manual, and observing a colleague use the device.

In a crisis, however, users must sometimes grab an unfamiliar device and use it based on their experience with similar devices. Because it is difficult for them to become experts in the use of all devices, medical workers must instead concentrate on the basics. They learn enough information to get their jobs done safely and effectively, then master the nuances of a device when they have time available.

Occasionally, medical workers' inexperience with a device creates problems or compromises patient safety. Often, the user's first response to such troubles is to speed dial the clinical engineering department for help or to riffle through overstuffed filing cabinets looking for the instruction manual, hoping it will solve the problem. At times like these, as well as during the initial process of learning to use a device, a readily accessible quick-reference guide may be the best solution.

A good quick-reference guide makes use of the so-called "20-80" or "key-task" rule, which states that users perform 20 percent of device-related tasks 80 percent of the time. A quick-reference guide should provide a limited amount of crucial information about key tasks—enough to accomplish the important jobs and nothing more (see chapter 7). This chapter describes the key elements of a quick-reference guide and how to create one.

Determining Content

When users are first trying to develop a general sense of how a device works, they may refer to a quick-reference guide. Therefore, a guide should present information in a way that helps build an accurate and effective mental model of device operations. With a good mental model in place, users can often anticipate how an unfamiliar function works.

A guide should also be designed to jog the user's memory about how to perform a specific task, outlining the basic steps without getting into detailed explanations. This content will give users comfort in the knowledge that they do not need to memorize every key press in order to operate the device. In this way a quick-reference guide works like an acrobat's safety net.

A quick-reference guide should not be a boiled-down version of a user manual. Instead, it should be a unique and important component of the overall user interface, developed according to specific design requirements that arise from its expected uses. Of course, for consistency's sake, the guide should use the same terminology and share certain visual elements with the device's on-line help, on-product labeling, and instruction manual.

User-interface design specialists, in consultation with marketing personnel and in-house clinical specialists, should perform a comprehensive analysis of key device tasks as a basis for selecting information to include in the guide. Depending on the product, the key tasks may have to do with primary device functions or maintaining/troubleshooting the device.

Most developers have a reasonably good sense of key device tasks, particularly when dealing with a mature technology. However, they should verify their list of tasks by talking to users about the information they would like to see in a quick-reference guide. In certain cases, when a product has been in use for some time without a published guide, users may have created their own guides. Naturally, such do-it-yourself guides are a good starting point for determining what should go into the official version.

Rapid Access to Information

Once the content of the guide is established, the next challenge is to provide quick access to the information. Most quick-reference guides are used in the trenches of daily medical practice—most users are in a hurry and under stress. They may be coping with several problems at once, and as a result may be more likely to forget things and make mistakes.

In the event users need help with a device-related task, they would probably first choose to talk to someone they think can help. The conversation would likely be quick and to the point, as in the following example:

User: I'm getting false apnea alarms.

Consultant: Check the ECG electrodes for good contact with the patient's skin.

User: They're all making good contact.

Consultant: Check the cable connections and lead wires.

User: They look OK, too.

Consultant: Check to see if the electrodes are too close to the apex of the heart, aorta, or liver.

In a similar manner a quick-reference guide might pose questions or conditions and provide short answers and explanations. Fundamentally, users should be able to pick up a guide, find the answers they are after in a matter of seconds, and get on with the task of using the device. If it takes users several minutes of flipping through and skimming pages to find what they're looking for, the guide has failed in its primary mission.

Features that enable users to retrieve information more rapidly also make documents more attractive and, in some cases, more expensive to produce. They include the following:

- A distinct typographical hierarchy that varies font style and size in a consistent and meaningful manner while avoiding excessive variation.

- Graphics that convey information in less space and with greater clarity than the use of words only.

- Section tabs (if there are a substantial number of pages and logical ways to divide the content).

- Headers, page numbers, and other symbols (such as icons) that help users identify important sections of a document.

- Borders or equivalent features to separate and demarcate related information.

- A high percentage of white space compared to text and graphics.

- The use of color to highlight certain information and to present graphics more realistically.

- The use of plain language.

- A short, logically organized table of contents.

- A short index (for longer guides).

- Cross-references to other pages in the guide, as well as to sections of the user manual.

- A binding that allows the document to stay open to a selected page.

Because rapid information retrieval is a high priority, quick-reference guides should be substantially shorter than user manuals. In fact, many people form their first impressions of the quality of a quick-reference guide based on its length, preferring shorter (one to five pages) to longer.

But given the varying complexity of devices and the ways they are used, it is impossible to specify an ideal length for a quick-reference guide. One guide might be limited to a two-sided, laminated card; another might need to be a 40-page, spiral-bound document.

Designers should not choose minimum length as their first priority. If not executed correctly, for example, a two-sided card may pack too much information into a small space, making important information hard to find and read. By comparison, colored section tabs, running headers, larger graphics, and more white space might make information in a longer document more accessible. Therefore, designers should first decide on essential content and then focus on good design principles.

However, developers must inevitably make hard choices about what to include and what to leave out of a guide. When there is too much to cover in a quick-reference guide designed for both new and experienced users, some manufacturers create another kind of document, perhaps titled "Getting Started," that focuses on device installation, setup, and initial use.

In order to attain maximum utility, a quick-reference guide must remain with its device, so manufacturers must take necessary measures to prevent it from being misplaced or lost. Some manufacturers build a special slot into the base or back of their product to hold the document, which may take the form of a rigid plastic card. Others attach a specially made pouch to their device, and still others tether the guide to the product with a chain or coiled cord. Each possible solution must be judged according to several criteria, including the following:

- Self-evidence to new users

- Accessibility during a task requiring intense concentration

- Susceptibility to damage or loss

- Interference with device operation

- Cost

Manufacturers' Viewpoints

Dennis Mattessich, a senior product manager with Datascope Corp. (Montvale, NJ), a manufacturer of patient-monitoring devices, considers the quick-reference guide important to the usability of his company's Passport portable bedside monitor (Figure 26.1).

> Our product is used in settings where people work in three shifts. Although we provide substantial training, we cannot count on the proper information getting to all possible users.

Figure 26.1. *Datascope's Passport portable bedside monitor.*

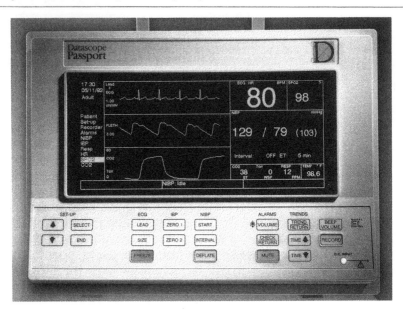

Courtesy of Datascope Corp., 14 Philips Parkway, Montvale, NJ 07645.

This makes an intuitive user interface and a good quick-reference guide that much more important.

Apparently, a good guide can also have a "halo effect" on the associated product.

A quality quick-reference guide says that a company cares about its customers' needs. If the customer sees we have put so much effort into developing a quality quick-reference guide, they assume we have put an equivalent effort into developing a product that will save them time and effort. So, in our view, you provide a quality guide because it is part of the total quality of the product.

According to Mattessich, Datascope put a lot of effort into the development of the Passport's quick-reference guide (what the company terms an abbreviated operator's guide). The original concept for the guide arose from the fact that early users improvised their own guides. The marketing staff established the necessary content, and an in-house industrial design group worked with marketing staff to produce the copy and the artwork. Datascope obtained preliminary feedback from its clinical education group, regional sales managers, and selected customers. Although this method of design review is not as structured as a usability test, it did reach out to representative users.

Explaining the rationale for the guide's design, Mattessich says,

> The monitor's control panel includes a key for every major function, rather than having the user select options from multilevel menus. In our guide, we reflect this design approach by showing a picture of each key and explaining what it does.

This design element allows users to scan the guide quickly to find out what the key they are interested in does. Furthermore, the guide gives the order of buttons for users to press to perform frequent tasks, such as controlling noninvasive blood pressure measurements and zeroing the invasive blood pressure lines (Figure 26.2).

The guide also contains extensive direction on how to place ECG leads on a patient's chest, including several pictographs (Figure 26.3). This level of detail might seem unusual, but Datascope's field research determined that many users were unsure about the correct placements of leads for adult versus neonatal cases. Therefore, the company took an extra step to ensure correct lead placement.

To make it easier to use, the guide is compact ($4^1/_2 \times 4^1/_2$ in.) and short (10 pages). Presently, it is attached to the monitor's built-in handle by a short chain, a compromise solution. Looking toward the next product generation, Mattessich envisions incorporating on-line help (which could be updated over the course of several software releases) as a possible replacement for the guide. However, it is not clear that on-line help will provide the same sense of immediacy and usability as a dedicated physical document.

Product developers at Ciba Corning (Medfield, MA), a manufacturer of blood chemistry analysis devices, also believe in the value of high-quality quick-reference guides. Some of their guides take the form of stand-up flip charts (with as many as 35 pages), which enable hands-off use (Figure 26.4). Concise text, simple drawings of device elements in a numbered sequence, and colored arrows highlighting the action to be

Figure 26.2. *The quick-reference guide for the Datascope patient monitor describes a series of buttons for users to push to complete frequently required tasks.*

> **1 Front Panel Controls**

Start

NON INVASIVE BLOOD PRESSURE
- Press START to initiate a measurement.

Interval
- Press INTERVAL to set or change an interval.

Deflate
- Press DEFLATE to stop measure- ment or interrupt a measurement cycle.

Zero 1

INVASIVE BLOOD PRESSURE
- Open the transducer stop-cock to atmosphere.

Zero 2
- Press and hold ZERO 1 or ZERO 2 for 1 second until IBP values go to zero.
- Turn stop-cock off back to atmosphere to begin monitoring.

Courtesy of Datascope Corp., 14 Philips Parkway, Montvale, NJ 07645.

---

taken ensure users rapid access to important information. Some pages of the guides have less text than normal and larger font sizes to facilitate viewing from a greater than normal distance. The company also matches a graphic with several sentences of explanatory text so that the guide serves the needs of people who learn better from pictures than from words.

Ciba Corning's quick-reference guides tend to run long, but they are much shorter than the user manuals provided with the products. Furthermore, the charts adhere to the first principle of quick-reference guides: The user should be able to obtain rapid access to basic and critical information.

Despite relatively low production runs (in the hundreds rather than the thousands), a factor that increases costs

Figure 26.3. *Pictographs in the Datascope patient monitor guide show optimal ECG lead placement and configuration.*

CONVENTIONAL LEAD PLACEMENT

Lead II:
Gives the best waveforms and is the most conventional method.
To achieve the best respiratory waveforms you may have to bring the Black and White electrodes down lower on the chest as illustrated below:

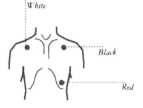

Conventional Electrode Position for Typical ECG Monitoring

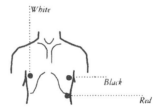

Conventional Electrode Position for Improved Respiratory Waveform Monitoring

Courtesy of Datascope Corp., 14 Philips Parkway, Montvale, NJ 07645.

substantially, Ciba Corning is committed to quality guides. They are typically made from heavy, laminated stock and include custom die cuts to form section tabs and a built-in easel.

Ciba Corning's Paula Hammett, a technical communications and usability specialist who now supervises the company's user-interface design efforts, has written several guides.

Our users say they depend on our guides and use them on a daily basis. They only use our instruction manuals when they have to look up the "gory details." The guides focus largely on

Figure 26.4. *A page from Ciba Corning's flip-chart quick-reference guide for its Express Plus random-access chemistry system.*

Preparing the System

 BIOHAZARD: For detailed precautions, refer to Appendix B, *Safety Summary,* in your *Express Plus Operator's Manual.*

This procedure assumes that your Express Plus and external printer (if installed) are on. If they are off, turn them on.

1. At the Startup screen, remove and empty the water bottle. Fill it with fresh deionized water; then install it in the system.

2. Remove and empty the waste bottle; then install it in the system.

3. Remove and discard the biohazard liner and its contents. Install a new biohazard liner; then install the cuvette waste drawer in the system.

4. Install new magazines of cuvettes in the cuvette feeder.

5. Press *ENTER.*

Reprinted with permission of Ciba Corning Diagnostics Corp.

critical tasks, such as troubleshooting and maintenance procedures that must be accomplished successfully to keep our analyzers working.

In the troubleshooting section of the guide (Figure 26.5), users locate the message in the left column that corresponds to the message on the device display and take the appropriate action listed in the right column.

If a lab technician cannot get an analyzer to work properly, the situation can be serious, because the devices are often used to analyze the blood of patients undergoing surgery, and the results of the tests can determine the course of an operation.

Figure 26.5. *Troubleshooting page from Ciba Corning's flip-chart guide.*

Digital Diluter Failure

1. Press F2 System Details; then look for the message next to Diluter.

2. Write down the message and any CPS code.

3. Identify the type of message and find the appropriate corrective action.

If the message is . . .	Then . . .
Cannot detect home position Requires Initializing	a. Press *MAIN MENU*; then press F7 Diagnostics/Maintenance.
	b. Press F2 Manual Operations; then press *ENTER*.
	c. Press F3 Prime Digital Diluter.
	d. If the message still shows, call Ciba Corning.
Syringe mechanism slipping	a. Clean the syringe and adjust the air gap.
	b. Press *MAIN MENU*; then press F7 Diagnostics/Maintenance.
	c. Press F2 Manual Operations; then press *ENTER*.
	d. Press F3 Prime Digital Diluter.
	e. If the message still shows, turn the system off; then turn it on.
	f. If the message still shows, call Ciba Corning.

Reprinted with permission of Ciba Corning Diagnostics Corp.

Hammett says,

An attractive guide is beneficial in two ways. First, it's going to be easier to read and find information. Second, it attracts attention, which is the first step in getting people to use it. And when they use it, they are less likely to make mistakes or get stuck on a task. An attractive guide also sends a marketing message: that the product will be easy to use.

This last message may seem to run counter to common sense, which suggests that a product that is easy to use does not need such an extensive quick-reference guide, or that a guide compensates for a poorly designed user interface. However, it is fair to assume that nobody is an instant expert at using a product. Like training wheels on a bicycle, a quick-reference guide may provide just the right level of assistance to people learning to use a product. Later on, its importance will naturally diminish, and it will offer value only when the user tries to perform an unfamiliar task.

As a professional who has campaigned hard for quality documentation, Hammett advises developers who are creating a quick reference guide to take full advantage of the skills of a technical communications specialist who is able to step back from the "gory details" and give users the information they really need to get their jobs done. She also advises companies to conduct usability tests of any user documentation.

Conclusion

Depending on the scope and complexity of user-device interactions, how much time users must spend learning to use a device, how often they use it, and how well the guide is designed, a quick-reference guide may be viewed as a critical device accessory or as a useless throwaway. Some types of devices may not need a first-rate guide, because they require very few user interactions and may include sufficient on-product labeling to provide the necessary operational guidance.

However, most people who use complex medical devices can benefit from a well-designed guide, which provides useful information and a measure of reassurance to users that they have a place to turn to if they need help. Furthermore, if the guide accomplishes nothing else, it serves as a symbol of the manufacturer's commitment to customers and to product usability.

Chapter

27

In Search of
Effective Warnings

When it comes to warnings, some people are believers and some are nonbelievers. Believers have faith that a well-designed warning will capture people's attention and lead them to behave properly and safely. They find fault with manufacturers who do not anticipate and warn against all hazardous conditions and behavior. By contrast, nonbelievers figure that most people ignore warnings and that placing warnings on a product is just a defensive response to a litigious environment. They argue that the recent proliferation of warnings addressing open and obvious hazards are a blight on society and, by creating a credibility gap, might actually reduce the attention paid to hazards. As with most things in life, the truth is likely to be found somewhere between these extreme viewpoints. Moderates recognize, for example, that only a few consumers are likely to notice and read thoroughly the multitude of warning labels found on a power saw and in its user manual. At the same time they would acknowledge that a sign reading "Sharks—Beach Closed" is quite effective at keeping people out of the water.

Regardless of one's viewpoint, warning labels are sure to be permanent fixtures on products and in user manuals.

Accordingly, manufacturers should strive to design warnings so that they communicate effectively. Abundant research has investigated the factors that contribute to or hinder the communication of warning information, most of it focused on devices capable of maiming or killing a person, such as power tools or car crushers. The findings, however, can be readily applied to products in the medical device industry.

Research on warnings indicates that designing an effective warning is a complex task; warnings should not simply be laminated or silk-screened distillations of some legal department memo. Legal departments are prone to create forbiddingly long lists of dos and don'ts that do not get read. A more effective approach is to combine the efforts of document design and usability specialists to produce warnings that grab the reader's attention and communicate quickly the vital information that can protect product users from hazardous situations.

According to warning-design specialists Lehto and Miller (1986, 5–6),

> The complexity of the warning issue arises from the [combined interaction] of possible users, products, [and] environments. . . . Each combination influences or is associated with the human's processing of information, upon which the ultimate consequences of a warning depend.

Accordingly, any conclusions from failure modes and effects analysis performed to define potential hazards are to some degree uncertain. In other words, it probably is impossible to predict exactly how an individual will use a product and interpret a warning. Investigators should perform sufficient research, however, so they are confident that they have identified the most important failure modes and effects.

Lehto and Miller (1986, 16) describe warning labels as

> stimuli [that] alert a user to the presence of a hazard, thereby triggering the processing of additional information regarding the nature, probability, and magnitude of the hazard. The additional information may be within the user's memory or may be provided by other sources external to the user.

Lehto and Miller also prescribe a process model for how warnings work. Their model, which is based heavily on communication theory, includes structural components (people, products, environments) and procedural components (activities that take place). The potential for such in-depth design analysis should give pause to product developers for whom warnings are an afterthought.

A State-of-the-Art Warning Label

Warnings are a subset of labeling; good labeling contains no distractions. (Parentheses, for example, are an automatic distraction.) Figure 27.1 shows a warning style consistent with guidance developed by the American National Standards Institute (ANSI). The warning incorporates the traditional signal word *danger*, sized and color coded to draw the attention of product users. Signal words such as *danger, warning,* and *caution* tend to function like symbols that are recognized and understood at a

---

Figure 27.1. *An effective warning message.*

Source: *Product Safety Sign and Label System*, FMC Corp., Santa Clara, CA. (FMC Corp. states that the words *high voltage* may not be necessary because the pictorial makes the meaning clear.

glance—much like the word *stop* on a stop sign. In other words, signal words produce an automatic initial response.

The balance of the warning contains a pictorial and some text. The pictorial has a simple appearance designed so that a majority of potential users will correctly interpret its meaning from only a brief glance. In this case the pictorial shows prohibited behavior and the text describes the hazard and method of avoidance. (Some warning messages, on the other hand, show proper, safe behavior.)

Overall, the warning in Figure 27.1 has a balanced, uncluttered look that encourages people to read it. The large amount of empty space makes the important information stand out. Even a quick glance is likely to communicate the intended message.

Common warning elements include signal words, text messages, and pictorials.

Signal Words

To promote greater consistency in product safety signs and signals, ANSI developed definitions for the following signal words:

- DANGER—indicates an imminently hazardous situation, which, if not avoided, will result in death or serious injury. This signal word is to be limited to the most extreme situations.

- WARNING—indicates a potentially hazardous situation, which, if not avoided, could result in death or serious injury.

- CAUTION—indicates a potentially hazardous situation, which, if not avoided, may result in minor or moderate injury. It may also be used to alert against unsafe practices.

Note: DANGER or WARNING should not be considered for property damage accidents unless personal injury risk appropriate to these levels is also involved. CAUTION is permitted for property-damage-only accidents (ANSI Z535.4-1991, 3).

As shown in Figure 27.2, each signal word is assigned a distinct color to make it recognizable as a symbol.

If a manufacturer uses ANSI standard definitions, the appropriate choice of a signal word may be clear, assuming that the hazard and consequences of using a particular device are well understood. The clear choice, however, may part with established industry conventions or conflict with the manufacturer's historical practice. As a result, changing from *danger* to

Figure 27.2. *Signal words for warnings; each has a distinctive color.*

White characters on red background

Black characters on orange background

Black characters on yellow background

Source: *Product Safety Sign and Label System*, FMC Corp., Santa Clara, CA.

warning for the sake of consistency with ANSI definitions could be interpreted as de-emphasizing the hazard, and possibly increasing legal vulnerability.

As an aside to the issue discussed above, experiments show that using a seemingly more-fear-arousing signal word such as *deadly, fatal,* and *lethal* will not necessarily increase the rate of compliance among trained professional users (as opposed to lay consumers). Furthermore, compliance among professionals does not seem to differ depending on whether the word *danger* or *warning is* used, although the word *warning* yields greater compliance than the word *caution* (DeTurk and Goldhaber 1989, 103–113).

Text Messages

Both prohibitive and permissive warnings can be effective, although individuals and population subgroups may respond better to one than the other. A well-known prohibitive warning is "Don't drink and drive." In another example a contact lens disinfecting unit includes in its instruction manual the following warning:

> If cracks appear in the disinfecting unit heat-well or on any other part of the unit, DO NOT USE as this may cause electrical shock.

This prohibitive warning includes a statement of consequence to reinforce the seriousness of the hazard and to explain the prohibition—an extra step to convince people to comply.

A familiar permissive warning, on the other hand, again from a campaign against drunk driving, is "Take the keys, call a cab, take a stand: friends don't let friends drive drunk." As another example the FDA requires that products incorporating ultraviolet generators include the permissive warning "Wear protective goggles during use."

Designers can use their experience and/or intuition to choose between a prohibitive and permissive approach. A more defensible approach, however, is to develop warnings employing each approach and conduct user testing.

When describing the consequence of a prohibited behavior, one must determine an appropriate threat level. In the case of the disinfecting unit, the threat is relatively mild ("this may cause electrical shock"). The threat level might be increased by stating "you could be electrocuted," a presumably stronger threat that appears on some handheld hair dryers. The literature on warnings contains considerable guidance on selecting appropriate threat levels (Evans, et al. 1970, 220–227).

Pictorials

There is an ongoing debate about the effectiveness of pictorials. They certainly assist communication with people who have low literacy skills or do not speak the pertinent language. Accordingly, warnings containing pictorials may be particularly beneficial for medical and diagnostic devices intended for in-home and multinational use. While a pictorial may increase understanding of a hazard, however, it is not certain that pictorials increase warning compliance among individuals who speak the language and have good literacy skills. Again, there is a wealth of literature addressing this issue (Friedman 1988, 507–515). Nevertheless, there is a trend toward greater use of pictorials in conjunction with text. The idea underlying this approach is that, because of its conspicuousness, a pictorial will draw an individual's attention to the warning message and communicate a portion of it, usually the prohibited behavior and the consequences, which is then reinforced by the text. Rarely does one find a warning that depends solely on a pictorial, perhaps because such designs require considerable artistic talent and it is easy to augment a pictorial with text.

Some good sample pictorials are presented in *Product Safety Sign and Label System* (FMC Corp. 1990), a workbook that delineates a step-by-step approach to pictorial development using standard symbols for people and body parts. The resulting pictorials, samples of which are shown in Figure 27.3, appear somewhat abstract—much like the symbols established for Olympic sports. These simple silhouettes are clearly visible in dim lighting and when viewed from afar. Some applications, however, may call for more-detailed depictions of people and tasks, such as one finds on airline safety cards (Figure 27.4).

In the interest of reliable pictorial interpretation, ANSI has established testing methods requiring developers to compile short descriptions of a given pictorial from a sample of at least 50 potential users. These descriptions are then sorted into four

Figure 27.3. *Sample pictorials. The warning message is clear even without accompanying text.*

Source: *Product Safety Sign and Label System*, FMC Corp., Santa Clara, CA.

Figure 27.4. *A good-quality pictorial used to communicate safety procedures to airline passengers.*

Courtesy of USAir; safety instructions for B737-200/300.

categories: correct, wrong, critical confusion (the user's interpretation provokes a dangerous response), and no answer. ANSI's acceptance criterion is

> 85% correct responses with a maximum of 5% critical confusions (ANSI Z535.3-1991, 20).

Design Considerations

Besides textbook recommendations and purist viewpoints on warning development, there are legal and regulatory issues to consider.

The Risk of Making Improvements

Paradoxically, manufacturers who take the initiative to improve their warnings may increase their vulnerability to a product liability claim based on failure to warn, because new and improved warnings seem to impugn the effectiveness of earlier versions. A plaintiff's attorney may argue in court:

Why would this company undertake a $200,000 program to improve its product's warnings unless the existing warnings were flawed?

This line of thinking could easily be reinforced by a "before and after" exhibit. This dilemma has led manufacturers in some industries to seek standard or "consensus" warnings, offering protection in numbers.

The Uncertain Need to Warn Expert Users

A special consideration is the possibility that a product is exempt from FDA requirements for adequate directions for use. According to the federal code, prescription devices do not necessarily have to be supplied with comprehensive instructions or warning statements, particularly when this information functions essentially as a reminder. The code states that such information may be omitted from the dispensing package if, but only if, the article is a device for which directions, hazards, warnings, and other information are commonly known to practitioners licensed by law to use the device (21 CFR 801.109).

Such exemptions may eliminate an overapplication of warnings on medical products for which users in fact fully understand hazards and methods of avoidance. Although an exemption eliminating a warning poses a fundamental shift of responsibility to the user, manufacturers still must decide whether it is reasonable to assume that all "practitioners licensed by law to use the device" will become and remain cognizant of all operating hazards. This is a tall order for medical workers who may rotate among several clinical sites and deal with dozens of devices, all of which may have warnings.

An Attorney's Viewpoint

Boston attorney Edward M. Swartz has strong opinions on the role of warnings in product safety, which he has shared in several books and on national television talk shows. In *Swartz Proof of Product Defect,* he states:

Warnings and instructions (labeling *in* FDA terminology) can be key in controlling and reducing the hazards of a product defect, whether the defect is retrospectively discovered or whether it was consciously incorporated into the design to improve function and/or efficiency. . . . (Swartz 1985, 575).

In short, Swartz is a believer and promoter of improved warnings. He views warnings as a secondary line of defense, however, stating:

> The first line of defense is always adequate design.

When asked for his general assessment of medical product warnings, Swartz responded critically:

> In certain instances, you see a considerable amount of overkill. Warnings will tell you about every possible hazard with no qualitative distinctions among those hazards. You get what amounts to a compendium of information that fails to communicate because it hasn't been designed to communicate. You get the feeling people [the manufacturers] are paying lip service, considering the poor quality of many warnings.

Swartz recommends that manufacturers make better use of experts in human factors, linguistics, psychology, and related fields in order to come up with better warnings. He views warnings as an integral part of an education process aimed at ensuring proper and safe use of medical products.

Swartz is concerned that the people who use medical products are not getting adequate training in their safe use. He cited interactions between salespeople and product users as a possible breakdown point in the education process. In his opinion,

> Salespeople are largely trained to push the product rather than to train others. We have no way of monitoring sales reps to see what information they are passing on to the users.

Swartz feels that heavy reliance on in-service-style training can lead to trouble later on when safety information provided through limited training is forgotten or recalled incorrectly. He advises manufacturers to reinforce safety practices taught through training with well-written instructions and good warnings.

To medical product designers, Swartz offers the axiom:

> It is always best to be reminded about the things we think we know cold.

When asked how he reconciles his criticism of warnings that practice overkill and the need for warnings that remind us of the familiar, Swartz dismissed the problem as

> a simple matter for design professionals, if the industry chooses to use them.

Swartz concurs with a human factors approach to warning design, which stresses user involvement.

> There is no more valid way to support a warning design effort than carefully controlled, safe user testing. Honest evaluation of warnings is what the industry needs.

He cautions manufacturers against conducting warning evaluations with people who work for the company or have special allegiances, however.

Swartz says that products with active patents have the best warnings.

> A patent makes a company much more forthcoming about potential hazards. [Such companies] are not concerned about frightening off buyers, because they are the only source [of the product]. Stronger competition often leads to consensus warnings in which everybody is providing the same safety information whether it is adequate or not. A competing product manufacturer is afraid to give too many threatening warnings, as it will otherwise make their product look more dangerous than [their competitors'].

Regarding the improvement of existing warnings and overcoming the problem of consensus warnings, Swartz said,

> Manufacturers have a moral, ethical, and legal obligation to constantly reassess their designs and warnings.

He considers it immoral for manufacturers to forgo improvements because they will make the installed base of products and warnings seem inadequate. Swartz advises that companies committed to product safety and to effective warnings will have a more defensible position in the event of a product liability claim.

Swartz likens the state of the medical product industry to that of the automobile industry several years back when it resisted certain safety devices such as seat belts and, later on, air bags. He expects the medical industry to embrace safety measures to a greater extent in the near future.

A Communication Expert's Viewpoint

Dr. Gerald Goldhaber, president of Goldhaber Research Associates (Amherst, NY) and professor in the department of communications at the State University of New York at Buffalo, conducts research on the effectiveness of warnings, serves frequently as an expert witness in product liability cases, and has worked with hospitals and doctors to develop various communication programs. He believes in the value of a well-designed warning for the right application, but feels that a large percentage of warnings are misapplied and, therefore, do not serve their intended purpose.

Goldhaber describes doctors as a "generally lousy audience for warnings."

> Doctors are some of the least likely people to pay attention to warnings since, from the day they begin medical school, they are told that they are the cream of the crop, that they know it all.

Apparently, the self-assuredness that comes with being a doctor reduces an individual's receptivity to warnings. Goldhaber's research suggests that a reduced level of awareness about hazards and their consequences leads to a reduced level of receptivity to warning messages and vice versa. He also finds that a professional's familiarity with a given product also makes her or him less receptive to a warning. In a recent article Goldhaber states:

> To the extent that a professional frequently works with a product and has not incurred an injury, or [has not] observed someone else injured from using the product, he/she will not be motivated to examine product safety information communicated in a warning (DeTurk and Goldhaber 1989, 103–113).

Like Swartz, Goldhaber is concerned about overwhelming product users with warning information.

> Giving users more warning information does not mean that they are better off.

He draws support from a statement attributed to Supreme Court Justice Warren Burger:

> If we warn about everything, we warn about nothing.

Goldhaber also thinks it is detrimental to remind people about the things they should know cold.

> If you get into the game of reminding people, you are going to create a mess. A lot of the time, you see lengthy warnings concocted by attorneys that nobody reads. Less important warning information dilutes the effect of the more important warning information.

Goldhaber says that designing an optimal warning calls for a structured design process.

Goldhaber's approach to warning development calls for defining areas of uncertainty via a hazards analysis and then selecting the appropriate media of communication. When the time comes to select the communication medium, he asks the question, "Who does the consumer trust most?" When communicating to doctors, he recommends the person-to-person communication that takes place at technical conferences and in-service presentations by salespeople as relatively effective approaches.

Still, Goldhaber recognizes the importance that warning labels may play in educating a new user about the most severe hazards and the legal protection that such a warning may offer the manufacturer. In fact, he anticipates a boom period in new product-specific design initiatives now that ANSI has approved an updated standard on warnings.

> The standards will be cited in every lawsuit in the future having to do with warnings. This will lead manufacturers to do a lot of rethinking and redesign.

On a related point, Goldhaber observed:

> Once you have your warning designed, you have to be reasonably satisfied that it works.

He condemns an approach that would have people simply look at a warning label and critique its effectiveness. Instead, he prefers an approach that exposes consumers to a warning in the course of their normal activities, to be followed by a survey in which the consumer is asked two questions: Did you notice the warning? Using your own words, what message does the warning convey? The first question determines whether the message is sufficiently conspicuous; the second, whether people understand the message and the means of hazard avoidance.

Toward Improved Warnings

The task of designing warnings can be vexing. The plethora of conflicting technical and legal issues, exacerbated by the uncertainties of communicating to diverse individuals, could hinder the development of product-specific warning messages. Yet, products will continue to be brought to market, and users need to be made aware of potential hazards. Therefore, even though experts may hold divergent viewpoints on when and how best to warn, developers need to take the design process seriously, making a significant investment of time and money. As Swartz suggests, a dedicated effort to design effective warnings is probably a company's best bet for protecting the consumer as well as the company.

References

21 CFR 801.109. *Code of Federal Regulations*, Food and Drugs 21 Parts 800 to 1299, revised as of April 1, 1990. Subpart D—Exemptions from Adequate Directions for Use, §801.109. Prescription devices.

ANSI Z535.3—1991. *Criteria for safety labels* (draft). New York: American National Standards Institute.

ANSI Z535.4—1991. *American national standard for product safety signs and labels* (draft). New York, American National Standards Institute.

DeTurk, M., and G. Goldhaber. 1989. Effectiveness of signal words in product warnings: Effects of familiarity and gender. *J Products Liability* 12:103–113.

Evans, R., R. Rozelle, T. Lasater, T. Dembroski, and B. Allen. 1970. Fear, arousal, persuasion, and actual versus implied behavioral change. *J Personality and Social Psychology* 16(2):220–227.

Friedmann, K. 1988. The effect of adding symbols to written warning labels on user behavior and recall. *Human Factors* 30(4):507–515.

Lehto, M., and J. Miller. 1986. *Fundamentals, warning—Vol. 1: Design, and evaluation methodologies*. Ann Arbor, MI: Fuller Technical Publications.

Product safety sign and label system. 1990. Santa Clara, CA: FMC Corp.

Swartz, E. 1985. *Swartz proof of product defect*. Rochester, NY: The Lawyers Cooperative Publishing Co.

Section

8

Special Topics

Chapter

28

The Usability Payoff: Case Study of a Syringe Pump

Many medical product developers hesitate to invest in usability. They want concrete reasons for spending limited R&D funds on it. They demand solid evidence of the usability payoff. Unfortunately, usability's exact benefits cannot be calculated. Even assuming it influences product purchase decisions, developing a precise formula for usability's return on investment is infeasible. This limitation is not unique to medical products.

Apple Computer has sold innumerable Macintosh computers (Macs) on the basis of the machine's ease of use. Its advertising has stressed the Mac's user-friendliness. While many other factors have certainly affected sales, the Mac's success is largely attributed to its usability. It has become a symbol of usability in the computer industry. The Mac's innovative user interface and a new awareness among consumers that computers can be easy to use have driven other computer manufacturers to improve their products' usability. Manufacturers have accepted Apple's customer satisfaction and strong product sales as evidence of a usability payoff and have increased their own investment in usability. Today, leading computer hardware and software developers employ user-interface design specialists or draw on consulting support for this expertise. In fact, the Computer Systems Technical Group has become the Human Factors Society's largest technical group, with over a thousand members.

The medical industry lacks a symbol of usability equivalent to the Macintosh, although many products meet users' needs and are highly regarded for their usability by special user groups. It stands to reason that some of these products would aptly demonstrate the usability payoff in user satisfaction and sales. This belief motivated the search for an exemplary product to serve as a case study of effective user-interface design.

Selecting one product for case study was complicated by the range in product complexity, the overtones of spotlighting one product versus a competing one, and the fact that no product is perfect (including the Macintosh). Limiting the domain to commonly used products of moderate technological and task complexity simplified the selection, yielding a syringe pump called INFUSO.R.™ (Bard MedSystems Division, C. R. Bard, Inc., North Reading, MA) (Figure 28.1). Distinguishing this product are its innovative user-interface design, its users' loyalty, and the fact that it performs a crucial function calling for timely and effective user interactions.

Primer on Syringe Pump Operation

A syringe pump provides a means for administering intravenous anesthetics. Primary users are anesthesiologists and nurse anesthetists. Users engage a syringe (containing the anesthetic drug) with the pump, which, in essence, supplants the user's thumb on the syringe plunger. This effects the downward movement of the plunger, thereby discharging the drug through an intravenous line and into the patient's bloodstream. The user specified via control settings the concentration of drug to be established in the patient's bloodstream (infusion rate). In addition to delivering the anesthetic at a continuous rate, the pump can be set to deliver a bolus injection: a volume of the drug injected at once.

Today, a growing family of intravenous drugs can be delivered via syringe pumps. Generally, these are of the fast-acting variety that comprise an intravenous approach to administering

Figure 28.1. *The INFUSO.R.™ syringe pump from C. R. Bard's Bard MedSystems Division.*

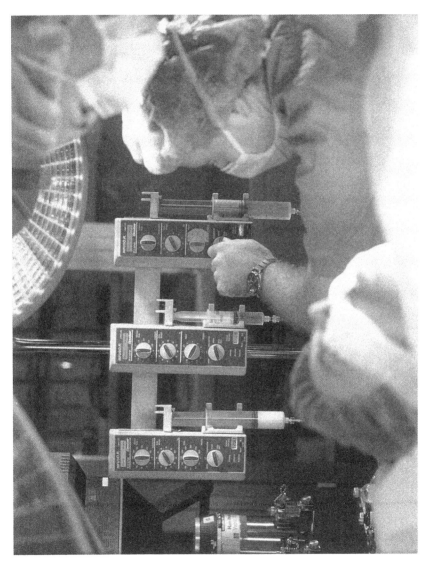

Courtesy Baxter Anesthesia Division.

anesthetics, replacing the more traditional approach of employing inhaled anesthetic gas. Fast-acting drugs have differing dose/response functions. Some become effective within 15–30 seconds; others over a period of several minutes. Therefore, both drug type and infusion rate (the amount of drug infused over time) influence the concentration of drug in a patient's bloodstream. The patient's body weight, which correlates to the total volume of blood in his or her body, also influences drug concentration. The number of interrelated factors make administering the appropriate dose of anesthetic potentially complex.

The User Interface of INFUSO.R.

INFUSO.R.™ is a compact, simple-looking product that many users consider easy to use. Its dominant feature is a removable, magnetic faceplate designated the Smart Label™. On first glance this component's primary purpose would appear to be labeling the drug in use. The balance of the faceplate includes scale markings for the positions of rotary knobs that set the infusion rate (defined as the resulting concentration of drug in the patient's bloodstream), patient body weight, and bolus. However, a comparison of several Smart Labels™ reveals that scale markings vary based on drug type. Moreover, faceplates are encoded magnetically to tell the syringe pump the drug type being infused. This information affects how the pump interprets control positions, which, in turn, affects the speed at which the syringe plunger is displaced downward. In essence, the Smart Label programs the pump. The user interface includes an additional control with permanent settings for pump operation mode (purging, off, stop/confirm/infuse, and bolus). Three LEDs labeled "infusing," "attention," and "bolusing" indicate pump operation mode. An LCD confirms Smart Label™ type and the syringe size the user has selected by means of a slide switch. Several design aspects contribute to the product's overall usability.

Ease of Learning

Rotary knobs provide an instant cue to users about how to adjust the control settings; you just turn them. Instructions for using the Smart Label™ are printed on the body of the pump and the Smart Label™, lessening the need to refer to a user manual. The controls perform dedicated functions, preventing the ambiguities that arise when a single control serves multiple purposes that change depending on operation mode.

Error Prevention

The drug name is color coded in accordance with ASTM standards for anesthetic drugs, facilitating instant recognition of the class of drug being infused and lessening the chance of infusing the wrong one. When the user attaches the faceplate to the pump, a small LCD below the faceplate displays a Smart Label™ code, such as L04. The user compares the code displayed to a code printed on the faceplate. A match confirms that the pump acknowledges the intended drug. To avoid incorrect dosing, drugs that come in varying concentrations have their own, unique Smart Label™.

Workload Reduction

The Smart Label™ enables the pump to be used for different drugs. This eliminates the time required to change syringe pumps during or after a given case. Scale markings for the infusion rate and bolus controls are positioned so that the typical setting is 12 o'clock (vertical). A glance at control positions tells the user whether the dosage is higher or lower than is typical, without the need to read the exact dosage. The pump calculates the proper infusion rate based on the control settings for desired drug concentration in the patient's bloodstream and for the patient's body weight, thus avoiding calculation errors and saving time. Earlier products required users to read special charts to determine an appropriate infusion rate. Switching between the infusion and bolusing modes is accomplished by the turn of a knob.

A User-Driven Design

The creativity of users who jury rigged an earlier pump (Bard's Alfentanil Infuser) spawned the innovative Smart Label™. Paul Lucas, a senior development engineering manager who worked on INFUSO.R.™, says,

> We had a good framework in the basic infusion device, but it wasn't flexible. It was programmed to infuse the drug alfentanil, but users started attaching homemade, paper labels to it.

Even though these did not change the pump's internal programming, they let people use the pump with other drugs. So, you could say people in the field started the movement toward the Smart Label concept. . . . We did not have to go through an extended research phase to determine how to make the product more usable. The users did the up-front work for us.

Blake Cerullo, product manager at Bard, says the Smart Label™ concept evolved from the work of Robert Kalayjian, M.D., of Long Beach Memorial Hospital (Long Beach, CA), who designed and sold a set of paper overlays called Infus-a-plates. These included scale markings and labels for several different drugs. The primary advantage of the Smart Labels™ over the Infus-a-plates is that they actually reprogram the pump. Control settings are typically programmed in the 12 o'clock position. Control positions are tuned to produce clinically suitable changes in infusion rate.

When Bard set out to improve its product by incorporating Smart Labels™, it sought to preserve the product's basic simplicity. Cerullo observes,

People working in the OR don't want to worry about their syringe pump. They don't have the time to be concerned with equipment during stressful moments.

Cerullo's view is echoed in Bard's choice to keep the product's user interface simple and intuitive to use. This explains why its latest design retains somewhat old-fashioned-looking rotary knobs in place of digital input devices such as the ubiquitous and often ambiguous arrow key. Cerullo acknowledges that the simplicity of the device may limit its perceived utility in settings such as the intensive care unit (ICU). He reports that Bard worked with anesthetic drug companies to determine the optimal settings for the infusion-rate control knob.

A User's Impressions

Jane Lowdon, M.D., is assistant professor of anesthesia at Emory University School of Medicine in Atlanta, Georgia. She practices anesthesia at Emory University Hospital, which maintains an inventory of 32 INFUsO.R.™ syringe pumps. Each of 15 operating rooms (excluding those for open-heart cases) is equipped with two. After operating the pumps for about two years, Lowdon regards them highly.

> The pump is very intuitive to use and the knobs are the main reason. One look at them and you know exactly what to do. They are self-explanatory.

She disdains products that do not put the controls up front where the user can both see them and figure out what they do. Specifically, products requiring several key presses and display changes to achieve a simple task frustrate Lowdon, and she reports that many of her colleagues feel the same way. She theorizes that

> people may like INFUSO.R.™ because they are used to rotating knobs. It's the way we do things in anesthesia. You turn lots of knobs, such as the vaporizer and nitrous oxide controls.

Lowdon also likes the way Smart Labels™ function.

> The plate that Bard has put on the pump gives reasonable ranges for the drug infusion scheme. It saves you the trouble of determining the normal range and prevents the confusion arising due to variations in the way infusion rates are specified. You don't have to worry about making mistakes converting from milligrams/kilogram/hour to micrograms/kilogram/minute.

She acknowledges some users' concern that INFUSO.R.™'s controls do not enable infinite adjustability for infusion rate or body weight. However,

> I don't feel I need any more precision than what the pump provides. I can titrate depth of anesthesia in many other ways, such as increasing the percentage of nitrous oxide.

She apparently views the limited precision provided by INFUSO.R.™ as a design attribute that prevents unnecessary decision making. On the negative side she believes INFUSO.R.™ and other syringe pumps require frequent surveillance—the kind that is normal in an operating room environment, but not in an ICU or recovery room. At Emory University Hospital syringe pumps are used only in the operating room. Lowdon believes an air detection capability and a way to visualize drug flow (as provided by drip chambers on some IVs) would improve the device's suitability for use outside the operating room. She cautions new users to confirm contact between the pump's driving mechanism and the syringe plunger before starting infusion because the device itself does not check for this condition.

Lowdon values ease of learning and gives INFUSO.R.™ high grades in this regard. She reports,

> We can teach residents coming into our anesthesia program how to use all the product's functions in five minutes, including cautions. In fact, you can really teach yourself how to use it.

In contrast,

> We have used another product that takes much longer to learn. In fact, I have had some people come back to me with questions after using it a year.

Lowdon warns medical product developers,

> If the product isn't quick and easy to set up, it will not get the anesthesia community's attention. Anesthesiologists and nurse anesthetists will not spend time going through a long setup procedure. If a syringe pump requires a lot of setup, they will just go back to using vaporizers. The fact that INFUSO.R.™ takes very little setup time has encouraged its acceptance.

The Payoff

The engineers and marketers at Bard are the first to admit that usability innovations in INFUSO.R.™ stemmed from its users. Bard benefited from the users' need for flexibility and usability. The company is now attuned to the usability payoff. It seeks a more formal and reliable method for collecting and addressing user needs in its designs. About INFUSO.R.™, Cerullo asserts,

> Ease of use is what makes it sell. There is a direct correlation between our strong sales and the product's ease of use. If we didn't have Smart Labels or an equivalent means for programming the pump, we wouldn't be seeing the same market success.

Publicizing Bard's experience developing and marketing INFUSO.R.™ may help others appreciate the usability payoff and increase investment in usability. Many products used today could set a comparably good example of user-interface design.

Chapter

29

Ergonomic Design of MIS Devices

Surgical practice is changing rapidly, in large part because of recent advances in video technology and surgical devices. Today, a growing number of surgeries, such as gall bladder removal (cholecystectomy), appendectomy, and hysterectomy, can be performed in a minimally invasive manner. With minimally invasive surgery (MIS) the surgeon's hands remain outside the patient's body. In place of making a large incision to access internal organs, surgeons make several smaller incisions through which specialized devices are inserted to perform the balance of the surgery. A miniature video camera connected to an endoscope (a telescope for looking inside the body) is the key to such procedures. When inserted into the abdominal cavity, for example, these devices provide the surgical team with a wide-angle, magnified view of the surgical field on a conventional video monitor. While watching a video image controlled by a surgical technician, scrub nurse, or another surgeon, a surgeon manipulates various rodlike devices. These devices are 1–2 ft in length and are equipped with specialized tips that grasp, clip, staple, cut, and cauterize tissue.

When practiced properly, MIS is impressive in terms of both the physical and mental skills exhibited by the doctors and the associated reduction in trauma to patients. However, a disturbing number of patient injuries, such as inadvertent laceration of the common bile duct, have been linked to design limitations of the equipment and to the ability of doctors to use it effectively. One basic problem appears to be immature technology, which requires users to adapt themselves to the technology, rather than the reverse. Another significant problem appears to be the limited availability of training opportunities and skill-development tools.

These problems could be addressed through more-extensive research on user-interface design topics, such as the influence of image quality on surgical effectiveness, the performance of three-dimensional manipulations based on two-dimensional views, the precise movement of tools with long handles, and oral communication among surgical team members.

Patient Benefits and Risks

When things go well with MIS, patients who required a month or more to recover from major surgery return to work much sooner. External wounds are closed with a few stitches and covered with a small dressing; scarring is minor compared to the long scars that have been the signature of conventional abdominal surgery. In this sense MIS may be considered patient friendly. Practitioners claim that a majority of patients who are presented with the option of laparoscopy (endoscopy performed in the abdomen) choose it over conventional surgery, despite its inherent risks.

There are, however, growing safety concerns. The June 14, 1992, edition of the *New York Times* reported:

> Since August 1990, at least seven patients have died and 185 others have suffered serious or life-threatening complications from the procedure, laparoscopic gall bladder surgery, at 99 of [New York State's] 242 hospitals. . . . An estimated 2% of patients undergoing laparoscopic gall bladder surgery in New York State have been injured because of surgeons' errors [according to the state's Health Department].

This injury rate compares poorly to an injury rate estimated at roughly 0.20 percent (1 in 500) for conventional gall bladder surgery.

Matching Technology to Human Capabilities

Both equipment manufacturers and medical experts agree that some doctors have difficulty adjusting to performing surgery by means of two-dimensional views of a patient's anatomy. Others may have difficulty coordinating their hand motions, which must be precise and which may appear reversed in the video monitor if the camera is directed toward rather than away from the surgeon. This can lead to surgical errors, such as placing a clip in the wrong location, dropping clips inside the wound, cutting the wrong tissue, cutting the correct tissue inaccurately, burning sensitive tissue by accidentally striking it with the hot tip of a laser-emitting device, or burning sensitive tissue because of inadvertent electrical conduction between the tissue and an electrocautery device.

In contrast to open procedures, in which the surgeon has direct visual and physical access to the patient's anatomy, MIS appears to leave little room for human error and less opportunity for correction. Many errors may be particularly hard to detect because the erroneous action is taken deliberately; doctors think they are taking the correct action when they are not, and they may not get immediate feedback to let them know that they have made a mistake.

While surgical difficulties suggest shortcomings in the skills of surgeons, they also suggest usability problems with evolving video and surgical devices. Of course, more training is a possible remedy—the State of New York concluded that practitioners may not be getting enough training before they operate independently on people. In response to this finding, it issued guidelines on laparoscopic surgery that require physicians to perform at least 15 procedures under supervision before a hospital can issue credentials that permit the physician to operate independently.

Physician Viewpoints

William Saye, M.D., founder and director of the Advance Laparoscopy Training Center (ALTC) in Marietta, Georgia, is a strong proponent of MIS. By 1992 his 40-person organization has trained a reported 8000 surgeons in laparoscopy over a three-year period. Surgeons attending ALTC's two-day courses must already be proficient at open surgery. Saye credits the boom in laparoscopy to innovations in video camera technology, which now provide surgeons with a well-lit, high-resolution, realistic color image on a video monitor.

> In the 1970s there was no video display. Surgeons looked directly through the laparoscope, held in one hand, to see what they were doing. As a result, they lost the use of one hand and, in effect, had to perform one-handed surgery. At the same time, nobody else in the operating room could see what the surgeon was doing, so he or she was working alone. Fortunately, laparoscopy is no longer a long-range, secret mission. By working with a video image, rather than looking directly through the laparoscope, surgeons have regained the use of both hands, and others can participate in a procedure.

In 1992 Saye estimated that 80 percent of all laparoscopic surgeries were being performed by one surgeon supported by a medical technician or scrub nurse. However, complicated procedures, such as a hysterectomy, may call for two surgeons to work together. Also, some surgeons prefer to work with a surgical partner. In a two-person team approach, one surgeon controls the camera view while the other manipulates the surgical instruments. Team members tend to talk a lot to each other in order to coordinate their actions and confirm anatomical landmarks.

Saye attributes a large percentage of reported patient injuries to the shortcomings of early technology. He says the early video cameras did not provide the same high resolution and bright image as today's cameras, which allow up to 15× magnification of the surgical site. In other words, doctors have not always had as good a view of what they were cutting as they do today. Saye also cites the development of better surgical instruments as a step toward reducing the rate of patient injury.

> An instrument called an Endo GIA [U.S. Surgical Corp., Norwalk, CT] has dramatically improved the speed and effectiveness with which we staple vessels and cut between them. It allows us to lay down two rows of three staples apiece and cut between the two rows, all in one [squeezing] action. With older devices, it might have taken 30 minutes to transect one blood vessel. The new device enables you to do the same thing in about 10 seconds.

Saye's organization teaches the fundamentals of laparo-scopy in the classroom, followed by three to five operations on pigs and a supervised operation on a human. "A physician can become proficient [at laparoscopy] after 10 cases or so," he says. However, Saye acknowledges differences in an individual's ability to learn and practice minimally invasive techniques effectively:

> The fact that someone is the best open surgeon does not neces-sarily mean he will be good at laparoscopy. That would be like saying that the best football player would be good at another sport, like baseball. You find that some people have a natural talent for the procedure, just like some people are good at using chopsticks.

Saye's remarks suggest that special tests (or exercises) could be developed to determine who has an aptitude for laparoscopy and then to build some of the necessary fundamental skills. Comparable tests are already used by the military to determine who has the mental and physical skills to become an effective pilot. For example, Korean aviators take tests designed to assess their ability to track a moving object displayed on a screen, react to visual and auditory input, recognize and remember items, estimate time and speed, and identify shapes. Would-be pilots must perform well on such tests before continuing their pilot training. (Park and Lee 1992, 189–204) However, it is not clear that surgeons would be receptive to such tests.

By 1992 Seymour DiMare, M.D., an experienced surgeon who attended an ALTC course in 1989, had performed more than 150 laparoscopies, working in partnership with Steven Margolis, M.D., at Emerson Hospital (Concord, MA). DiMare says the critical hurdle for newcomers to laparoscopy, or other types of MIS such as arthroscopy (surgery within a joint) and pelvis-copy (surgery within the pelvis), is working in two dimensions.

> Picture driving your car while keeping one eye closed. You lose depth perception, so you have to form a mental picture of where you are. It's the same situation when you switch from a direct view of the surgical field to a video image. Some people are able to make the adjustment because they have an innate sense for spatial relationships.

DiMare feels that the key to working with two-dimensional views is to keep the camera view steady, use the shadow cast by inserted instruments to judge distance, and work as a team to accurately identify the internal structures.

Despite their personal comfort working with two-dimensional views of the surgical field, DiMare and Saye both look forward to the widespread use of stereoscopic video, a demonstrated technology that has yet to be placed into general use. Saye feels that a stereoscopic view of the surgical site, provided through some type of headset (e.g., goggles), will dramatically increase the ease and efficiency of MIS and will very likely lead to a host of new applications.

DiMare, who had performed open gall bladder surgery for more than 20 years before switching to laparoscopy for most cases, feels there is still room for more-advanced training tools. For example, he would welcome the development of computer-based models for laparoscopic procedures, which would give doctors greater opportunity to practice their techniques.

While computer-based training will not supplant hands-on training, its potential seems great. After all, there are computer-based training systems for operating nuclear power plants and flying aircraft—both of which are complex, dynamic systems that must function properly in order to avoid placing lives at risk. Three-dimensional modeling software, available on engineering workstations, is surely up to the task of producing the visual fidelity necessary to depict anatomical structures. In fact, there has been progress in this area of modeling. Also, human factors and medical specialists working together could probably produce training exercises suited to building and maintaining fundamental skills—skills that may lapse unless a surgeon performs an endoscopic procedure with sufficient frequency. Developing computer-based training systems is costly, however. Therefore, apart from major technological barriers, the availability of advanced training approaches that do not rely on the use of living patients or animals may be a matter of economics.

Developing User-Oriented Designs

Ray Ogle is vice president of product development at Ethicon Endo-Surgery (Cincinnati, OH), a manufacturer of endoscopic surgical instruments.

> Ethicon's approach to equipment design emphasizes getting users [i.e., surgeons] involved early and throughout the development process. In fact, we have several surgeons on staff who specialize in minimally invasive surgery.

According to Ogle, Ethicon sets up so-called procedural teams that include engineers, industrial designers, and surgeons to focus on a particular class of operation, such as hernia repair (see Figure 29.1).

> Early on, a team will develop a [surgical] procedure description that is independent of design. The procedure-based description includes in great detail all of the mental and physical tasks that the surgeon must perform. Then we follow a highly iterative design process that includes concept development, CAD/CAM-based design and prototyping, and veterinary and clinical testing.

In addition to putting surgeons on the development team, Ethicon's engineering and design staff members receive an extensive clinical orientation. They attend three- to four-day anatomy courses, taught by on-staff surgeons, that focus on a procedure of interest to Ethicon. They work closely with surgeons and observe an average of 20 surgeries (open and minimally

Figure 29.1. *Ethicon's endoscopic instruments are developed by teams that unite surgeons with clinically trained engineers and designers.*

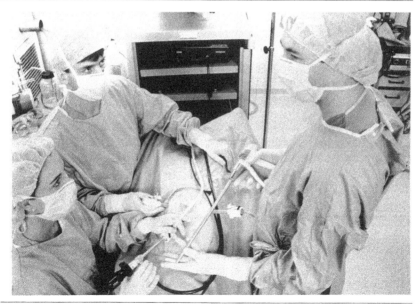

Photo courtesy of the Industrial Design Department of Ethicon Endo-Surgery, a Johnson & Johnson company.

invasive) at hospitals throughout the United States. According to Ogle, this rigorous clinical training improves team communication and sensitivity to usability issues.

Ogle feels the usability of one of Ethicon's latest endoscopic clip appliers reflects the company's investment in staff training and user involvement.

> Our research showed that doctors preferred to apply clips one-handed. This meant designing a device that facilitated a secure grip on the clips [prior to clamping], proper orientation of the clipping device's jaws, and secure attachment of the clip to the appropriate vessel.

To accomplish this goal, Ethicon's industrial designers, who are trained in human factors, performed analyses of hand size and strength to guide design development. The final product is a disposable device that automatically feeds up to 20 clips through a shaft 10 mm (0.4 in.) in diameter. Doctors use one hand to maneuver and rotate the clipping device into the proper position and then securely apply a clip. Reflecting on the product development effort, Ogle says,

> Producing such a precise surgical instrument at a reasonable price that also met our usability goals required us to push the technology envelope. It took many iterations, working with our CAD/CAM equipment and performing user testing in our laboratories, to arrive at the final design.

Recent and Near-Term Innovation

Clearly, the race is on to produce better imaging and surgical devices. Surgeons will be receptive to devices that require fewer steps, make things go faster, and reduce the opportunity for error, thereby improving patient safety.

John Dawoodjee is associate director of technical support at Karl Storz Endoscopy America, Inc. (Culver City, CA), a subsidiary of a German firm that manufactures medical video systems, endoscopes, and associated instruments. He feels that Karl Storz has made a commitment to usability that is reflected in several innovations. For example, the company's cameras include an automatic exposure feature that eliminates the need for the user to adjust constantly the attenuator on the light source.

> Our exclusive autoadjust feature is particularly valuable to doctors during operations on the ovaries, which have a whitish appearance and reflect a lot of light.

The company continues to use rod lenses in its cameras, rather than replace them with electronic sensors, so that doctors receive a true-color image to help differentiate anatomical features according to flesh tone. Storz's camera also has a manual zoom capability so that doctors can adjust the field of view to the desired level of magnification.

Like other companies, Karl Storz has upgraded to so-called three-chip cameras, which employ charge-coupled device technology to give physicians a better-quality image—one with less distortion and more lines of resolution. According to Dawoodjee,

> Whereas a normal television picture has 230 lines of resolution, and a standard, single-chip camera connected to a medical-grade monitor has 420 lines of resolution, Karl Storz's three-chip camera connected to a medical-grade video monitor approaches 700 lines of resolution.

Dawoodjee cites his company's laparoflator (a type of insufflator used in laparoscopy) as another advance in usability.

> The laparoflator is designed to automatically maintain a set volume and pressure of CO_2 in the patient's abdomen. This saves the doctor from constantly having to adjust the degree of insufflation.

The device has buttons for digitally setting the pressure and volume levels, as well as built-in alarms for these variables. Dawoodjee says the company has also redesigned its needle-holding devices used for suturing so that they have a more comfortable grip and so that the surgeon has increased manual control during the endoscopic suturing process.

Dawoodjee expects that endoscopy technology will continue its rapid evolutionary pace. He anticipates a new generation of surgical instruments that will make it easier to handle tissue removal, such as motorized devices that cut tissue into small pieces for removal through a cannula. He expects the industry to adopt image-enhancement software, like the software used by NASA to enhance the images of astronomical features. He also expects progress to be made on stereoscopic imaging, but he thinks that the first generation of devices will not provide the same high resolution as today's three-chip cameras, largely because [in 1992] it is necessary to use a pair of single-chip cameras (which produce fewer lines of resolution) to achieve the three-dimensional effect.

Another area for near-term improvement is workstation design. Doctors complain that today's equipment is frequently a miscellaneous collection of devices, manufactured by numerous vendors, that lacks physical integration. The penalty for such a haphazard approach is poor workspace ergonomics. For example, doctors often get stiff necks from the head twisting required to look at monitors, which are usually placed on the shelves of a rolling cart. Perhaps in the near future we will see endoscopic workstations that reflect the refinement of the latest generation of anesthesia monitors—boom-mounted displays, integrated controls, hierarchical alarms, ample table and drawer space for various accessories, and a small footprint (amount of floor space required). As is the case with some anesthesia workstations, integration of endoscopic workstations may require cooperation and standardization among manufacturers.

Long-Term Innovation

The user interface to minimally invasive surgical devices could undergo dramatic change over the next decade. Richard Satava, M.D., and Philip Green, researchers at the U.S. Army's Fort Ord (Monterey, CA) and at Stanford Research Institute International (Palo Alto, CA), respectively, have explored the application of telepresence, or so-called virtual reality techniques, to MIS. Although the main focus of their work has been to improve the safety and effectiveness of standard surgical procedures, they think that surgeons may someday operate on patients without even being at the same location as the patient (e.g., by interacting with a computer-augmented, stereoscopic image of a patient's body).

Satava and Green regard early 1990s techniques as primitive compared to what the future holds. They compare laparoscopic techniques introduced in 1989 as

> somewhat analogous to using sharp sticks in a bag. The [video] image is restricted and must be controlled by one of the surgeons. The instruments are not articulated; rather, they work like a fulcrum through the body wall. There is a [minimum] of direct feedback, not a true sensory input (Satava and Green 1991).

The researchers' concept is

> to create a defined space inside the body, like a miniature Astrodome, with the video camera suspended from the top of the dome (or umbilicus) and the surgical instruments inserted through the dome. Internal "vision" is provided by an internal stereoscopic camera, which is stabilized by external supports [connected] to the operating table. Camera positioning is regulated by either voice commands or by a head-mounted device on the surgeon (as his head moves, the camera moves). Four or five entrance ports are inserted into the body wall and stabilized to the side of the operating table; through these ports various surgical instruments are introduced. These [computer-controlled] instruments are articulated, giving more degrees of freedom than current laparoscopic instruments. When the operator controls are grasped, the remote manipulators appear in the virtual work space and move precisely as if they were in the [surgeon's] hands (Satava and Green 1991).

Is telepresence the logical extension of today's laparoscopic methods? ALTC's Dr. Saye remains open-minded but thinks computer-based tools will not be able to provide the control and sensory inputs necessary to address the subtle aspects of human anatomy and surgical procedures. Nonetheless, such research, focusing on the human factors limitations of today's laparoscopic techniques—principally visual feedback and precise manipulation of instrumentation—should pay off in the form of enhancements to current equipment and training methods.

Telepresence research may already be pointing the way toward improved methods for assessing surgical skill levels. Reporting on the precision of the remote manipulations achieved by their telepresence system, Satava and Green state,

> In defined tasks, a 3-mm rod can be threaded through a 4-mm washer without touching the sides 10 out of 10 times; the tip of a scalpel blade can touch the exact tip of a pin on 25 consecutive tries, and a grape held in an assistant's hand can be carved into 1-mm slices by a scalpel.

This type of testing might well be adapted to assess the skills of physicians performing established endoscopic procedures.

Conclusion

Minimally invasive surgery is awe inspiring when one considers the kinds of complex surgical procedures that can be performed by hands that remain outside the patient's body. The new surgical methods, regarded by many as the most important development in surgical practice in several decades, even take on a surreal quality, due in part to the use of video displays as a primary feedback mechanism. The glamour surrounding MIS diminishes somewhat, however, when one considers the relatively high rate of patient injury associated with minimally invasive procedures such as laparoscopy. Some suggest that a high percentage of the injuries can be attributed to physicians' inexperience with the new surgical methods and inadequate supervision. The situation is analogous to accidents resulting from pilots soloing in a new class of airplane before they gain sufficient experience in the procedures of taking off and landing. Therefore, New York State's strong action, establishing guidelines for the minimum number of surgeries that must be performed under supervision, seems to be a step in the right direction.

However, it is not just the surgeons who are responsible for the current rate of patient injury—medical device manufacturers are also responsible. Clearly, today's technology pushes users to the limits of their ability and, in effect, induces human error. For example, some surgeons experienced and proficient at open surgery have substantial difficulties working with a two-dimensional video image of an internal organ. Nonetheless, the same surgeons will probably persevere with their plans to perform MIS; there are marketplace pressures to do so. Perhaps manufacturers can alleviate some of the problems by developing new training techniques and certification exercises to accompany their products in order to ensure that users are properly prepared. Motives for manufacturers to move on this front include injury prevention and protection from product-liability actions. When comparing the training needs of surgeons to those of pilots, it is significant to note that airlines spend heavily to train pilots on advanced simulators. The simulators, which are able to generate a wide range of flying experiences, allow pilots to crash the airplane repeatedly until they get it right. Medical schools might consider emulating the airlines' policies.

Manufacturers should also pay greater attention to the physical and perceptual challenges created by the new technology and should set goals for improved usability. Judging by progress made in other industries (e.g., aviation, nuclear power, consumer software), better user interfaces should pay off in reduced human error. If such progress occurs, there should be fewer patient injuries, quelling current safety concerns.

References

Park, K., and S. Lee. 1992. A computer-aided aptitude test for predicting flight performance of trainees. *Human Factors* 34(2):189–204.

Satava, R., and P. Green. 1991. *Telepresence surgery, phase I: Basic concept and design.* Presented to the Virtual Reality Conference of the Education Foundation of the Data Processing Management Association, October, in Washington, DC.

Resources

Resources

Extensive resources are available to medical product developers who are concerned with improving the usability of their products. These resources fit into the following categories:

- Books
- Periodicals, proceedings, booklets, and newsletters
- Professional societies

Specific publications and organizations are listed below according these categories. The listings exclude the mainstream medical publications and organizations that may be identified by searching national databases, by contacting large medical libraries, and by obtaining a listing of government agencies.

Books

Bailey, R. 1982. *Human performance engineering: A guide for system designers.* Englewood Cliffs, NJ: Prentice Hall, Inc.

Bass, J., and D. Prasun. 1993. *User interface software.* New York: John Wiley and Sons, Inc.

Bogner, S., ed. 1994. *Human error in medicine.* Hillsdale, NJ: Lawrence Erlbaum Associates, Inc.

Brown, C. M. 1988. *Human-computer interface guidelines*. Norwood, NJ: Ablex.

Bullinger, H., and R. Gunzenhauser, eds. 1988. *Software ergonomics: Advances and applications*. New York: Halsted Press.

Casey, S. 1993. *Set phasers on stun and other true tales of design, technology, and human error*. Santa Barbara, CA: Aegean Publishing Company.

Cushman, W., and D. Rosenberg. 1991. *Human factors in product design, Advances in human factors/ergonomics*, Vol. 14. New York: Elsevier Science Publishers B.V.

Diffrient, N., A. R. Tilley, and D. Harmon. 1981. *Humanscale 4/5/6*. Cambridge, MA: The MIT Press.

Dreyfuss, H. 1967. *Designing for people*. New York: Grossman Publishers: A Division of Viking Press.

Dumas, J. 1988. *Designing user interfaces for software*. Englewood Cliffs, NJ: Prentice Hall.

Dumas, J., and J. Redish. 1993. *A practical guide to usability testing*. Norwood, NJ: Ablex Publishing Corporation.

Edmonds, E., ed. 1992. *The separable user interface*. New York: Academic Press, Inc.

Ehrich, R., and R. Williges, eds. 1986. *Human-computer dialog design, Advances in human factors/ergonomics*, Vol. 2. New York: Elsevier Science Publishers B.V.

Golman, A., and S. McDonald. 1987. *The group depth interview: Principles and practice*. Englewood Cliffs, NJ: Prentice Hall, Inc.

Hix, D., and R. Hartson. 1993. *Developing user interfaces: Ensuring usability through product & process*. New York: John Wiley & Sons, Inc.

Hix, D., and R. Hartson. 1988. *Advances in human-computer interaction—Volume 2*. Norwood, NJ: Ablex Publishing Corporation.

Horton, W. 1990. *Designing and writing online documentation*. New York: John Wiley and Sons, Inc.

Kearsley, G. 1988. *Online help systems—Design and implementation*. Norwood, NJ: Ablex Publishing Corp.

Klemmer, E., ed. 1989. *Ergonomics: Harness the power of human factors in your business*. Norwood, NJ: Ablex Publishing Corporation.

Larson, J. 1992. *Interactive software: Tools for building interactive user interfaces*. Englewood Cliffs, NJ: Yourdon Press.

Lorenz, C. 1990. *The design dimension: The new competitive weapon for product strategy & global marketing*. Cambridge, MA: Basil Blackwell, Inc.

Marcus, A. 1992. *Graphic design for electronic documents and user interfaces*. New York: ACM Press.

Martin, C. 1988. *User-centered requirements analysis*. Englewood Cliffs, NJ: Prentice Hall, Inc.

Mayhew, D. 1992. *Principles and guidelines in software user interface design*. Englewood Cliffs, NJ: Prentice Hall, Inc.

McCormick, E., and M. Sanders. 1982. *Human factors in engineering and design*, 5th ed. New York: McGraw-Hill Book Company.

Nielsen, J. 1993. *Usability engineering*. Boston: Academic Press, Inc.

Norman, D. 1988. *The design of everyday things*. New York: Basic Books, Inc.

Powell, J. 1990. *Designing user interfaces*. San Marcos, CA: Microtrend Books, Slawson Communications, Inc.

Putz-Anderson, V. 1988. *Cumulative trauma disorders*. New York: Taylor & Francis Inc.

Rubinstein, R., and H. Hersh. 1984. *The human factor: Designing computer systems for people*. Burlington, MA: Digital Press.

Salvendy, G., ed. 1987. *Handbook of human factors*. New York: John Wiley & Sons, Inc.

Schneiderman, B. 1992. *Designing the user interface: Strategies for effective human-computer interaction*, 2nd ed. Reading, MA: Addison-Wesley Publishing Company, Inc.

Smith S., and J. Mosier. 1986. *Guidelines for designing uuser interface software.* Technical Report ESD-TR-86-278. Bedford, MA: MITRE Corporation.

Stanton, N., ed. 1994. *Human factors in alarm design.* Bristol, PA: Taylor & Francis.

Tognazzini, B. 1992. *Tog on interface.* Reading, MA: Addison-Wesley Publishing Company, Inc.

Tufte, E. 1990. *Envisioning information.* Cheshire, CT: Graphics Press.

Tufte, E. 1983. *The visual design of quantitative information.* Cheshire, CT: Graphics Press.

Van Cott, H., and R. Kinkade. 1972. *Human engineering guide to equipment design.* Washington, DC: U.S. Government Printing Office.

Vassiliou, Y. 1984. *Human factors and interactive computer systems.* Norwood, NJ: Ablex Publishing Corporation.

Wiklund, M., ed. 1994. *Usability in practice—How companies develop user-friendly products.* Cambridge, MA: Academic Press.

Woodson, W., ed. 1981. *Human factors design handbook.* New York: McGraw-Hill Book Company.

Periodicals, Proceedings, Booklets, and Newsletters

ANSI/AAMI HE48-1993. *Human factors engineering guidelines and preferred practices for the design of medical devices.* Arlington, VA: Association for the Advancement of Medical Instrumentation.

ANSI/HFS 100. 1988. *American national standard for human factors engineering of visual display terminal workstations.* Santa Monica, CA: Human Factors and Ergonomics Society.

ANSI Z535.1. 1991. *Safety color code.* New York: American National Standards Institute.

ANSI Z535.3. 1991. *Criteria for safety symbols.* New York: American National Standards Institute.

ANSI Z535.4. 1991. *Product safety signs and labels.* New York: American National Standards Institute.

Applied ergonomics. Oxford, England: Butterworth-Heinemann Ltd.

Before and after: How to design cool stuff. Roseville, CA: John McWade.

Behaviour & information technology. Bristol, PA: Taylor & Francis.

Common ground: The newsletter of usability professionals. Usability Professionals Association.

Communications of the ACM. New York: Association for Computing Machinery.

Computers in human behavior. Elmsford, NY: Pergamon Press Inc.

Designer's handbook: Medical electronics. Santa Monica, CA: Canon Communications, Inc.

Ergonomics. Bristol, PA: Taylor & Francis.

Ergonomics in design. Santa Monica, CA: Human Factors and Ergonomics Society.

Human-computer interaction. Hillsdale, NJ: Lawrence Erlbaum Associates.

Human factors and ergonomics society bulletin. Santa Monica, CA: Human Factors and Ergonomics Society.

IEEE computer graphics and applications. Los Alamitos, CA: The Institute of Electrical and Electronics Engineers.

I.D. Magazine. New York: Magazine Publications.

Innovation. Great Falls, VA: Industrial Designers Society of America.

Interacting with computers. Oxford, England: Butterworth-Heinemann Ltd.

Journal of the human factors and ergonomics society. Santa Monica, CA: Human Factors and Ergonomics Society.

Proceedings of the ACM special interest group on computer-human interaction. Baltimore, MD: Association for Computing Machinery.

Proceedings of the human factors and ergonomics society. Santa Monica, CA: Human Factors and Ergonomics Society.

Proceedings of the national computer graphics association. Fairfax, VA: National Computer Graphics Association.

Proceedings of the society for information display. Playa Del Rey, CA: Society for Information Display.

Proceedings of the society for technical communication. Arlington, VA: Society for Technical Communication.

Proceedings of INTERACT. New York: Elsevier Science Publishing Company B.V.

Product safety label handbook. 1981. Trafford, PA: Westinghouse Printing Division.

Product safety sign and label system. 1985. Santa Clara, CA: FMC Corp.

SIGGRAPH (Special Interest Group on Computer Graphics) Bulletin. New York: Association for Computing Machinery.

Professional Societies

The professional societies listed below support the interests and needs of those involved in user-interface design. Where a society has no central office, the address of one of the society's organizers or officers is given.

Association for the Advancement of Medical Instrumentation
3330 Washington Boulevard, Suite 400
Arlington, VA 22201-4598

Ergonomics Society
Rod Graves, Secretary
University of Technology
Loughborough, LEIC LE11 3TU
England

Human Factors and Ergonomics Society (HFES)
P.O. Box 1369
Santa Monica, CA 90406

IEEE Systems, Man, and Cybernetics Society
345 East 47th Street
New York, NY 10017

Industrial Designers Society of America (IDSA)
1142-E Walker Road
Great Falls, Virginia 22066

Society for Information Display
8055 W. Manchester Avenue
Suite 615
Playa del Rey, CA 90293

Society for Technical Communication (STC)
Suite 904
901 North Stuart Street
Arlington, VA 22203

Special Interest Group on Computer and Human Interaction
(SIGCHI)
Association for Computing Machinery
P.O. Box 12115
Church Street Station
New York, NY 10249

Usability Professionals Association (UPA)
Janice James, Chairperson
American Airlines / STIN
P.O. Box 619616 MD 4230
DFW Airport, TX 75261-9616

To obtain a more extensive listing of societies, particularly those established outside the United States, refer to:

Pelsma, K., ed. 1987. *Ergonomics sourcebook*. Lawrence, KS: The Report Store.

Index

Name Index

Aaron Marcus + Associates, Inc., 127, 159
Abramovich, Abe, 32–33
Adleman, George, 237
Advance Laparoscopy Training Center
        (ALTC), 349, 351, 357
Albers, J., 146, 153
Allegiant Technologies, 248
Allen, B., 336
Altia® Design, 13, 249, 256
American Institutes for Research, 36, 289
American National Can, 218
American National Standards Institute
        (ANSI), 31, 116, 160, 163, 222, 225,
        325, 326–327, 329, 330, 335, 366, 367
American Society for Testing and
        Materials (ASTM), 32, 53, 144, 149,
        151, 169, 283, 343
Anesthesia Patient Safety Foundation, 62
Apple Computer, 126, 128, 248, 255, 339
Applied Science Laboratories, 84
Armstrong, T., 187, 195
Arnaut, Lynn, 224, 228, 234, 235, 239
Arthur, Frank, 218–219

Association for Computing Machinery,
        4, 28, 31, 138, 367, 368, 369
Association for the Advancement of
        Medical Instrumentation, 31,
        115–124, 148–150, 152, 366, 368
Asymmetrix, 249
AutoCAD, 107–108

Badler, Norman, 110
Bailey, R., 363
Banks, Seth, 161
Bardagjy, J., 210
Bard, Inc., C. R., 340, 341, 343, 344, 346
Barr, Robin, 176, 177, 178
Bass, J., 363
Baxter Laboratories, 341
Bennet, K., 135, 139
Bennett, J., 97, 102
Berkowitz, J., 182, 183
Bias, R., 35, 43
Biomechanics Corporation of America
        (BCA), 107–109
Blattner, M., 159, 164, 170–171, 172

Subject Index